ARTISTS, CRAFTSMEN, AND TECHNOCRATS

THE DREAMS, REALITIES, AND ILLUSIONS OF LEADERSHIP

SECOND PAPERBACK EDITION

PATRICIA PITCHER

Stoddart

Second paperback edition published in 1997 by
Stoddart Publishing Co. Limited
34 Lesmill Road
Toronto, Canada
M3B 2T6
Tel.(416) 445-3333
Fax (416) 445-5967

01 00 99 98 97 1 2 3 4 5

First published in hardcover by Stoddart Publishing in 1995
Published in paperback in 1996

Cataloguing in Publication Data is available from the National Library of Canada.

ISBN: 0-7737-5854-2

Cover design: Bill Douglas @ the Bang
Printed and bound in Canada

Stoddart Publishing gratefully acknowledges the support of the Canada Council and the Ontario Arts Council in the development of writing and publishing in Canada.

Contents

Foreword

I had the pleasure of supervising this work in its original form as a doctoral thesis. So I am perfectly biased. I know it is a critically important piece of work, tackled with a sophistication and elegance that is rare in the management literature. If the message of this book can be taken to heart by people in important places, our organizations will become entirely different and, in my opinion, much more effective places.

A leader has to be one of two things: He either has to be a brilliant visionary himself, a truly creative strategist, in which case he can do what he likes and get away with it; or else she has to be a true empowerer, who can bring out the best in others. Managers who are neither can be deadly in organizations that need energy and change. In this book, Pat calls the first *Artists* and the second *Craftsmen,* the third *Technocrats.* She shows in a pointed, deep study of a large financial institution how the Technocrats killed what the Artists built and the Craftsmen protected.

We have to understand these different management styles and what each can do to organizations. If you care about what's happening to business and other organizations in the Western world, this is a book you will want to read and cherish.

Henry Mintzberg
McGill University
Montreal

Introduction

About Sows' Ears and Silk Purses

> *Still more do I regret the failure to convey the sense of organization, the dramatic and aesthetic feeling that surpasses the possibilities of exposition, which derives from the intimate, habitual interested experience. It is evident that many lack an interest in the science of organizing, not perceiving the significant elements. They miss the structure of the symphony, the art of its composition, and the skill of its execution, because they cannot hear the tones.*
>
> Chester Barnard, *The Functions of the Executive*[1]

This book, and the 8-year study on which it is based, is the product, as are all such works, of a personal odyssey. After a 15-year career in the private sector, I had decided to do a doctorate in management and, like most practitioners, really hadn't the faintest idea what the academics said about managing. Being exposed to this academic viewpoint for the first time, in 1986, was a shock. It sure missed the tone of organization. With rare but welcome exceptions, there were no real people in most of the management literature. There were recipes. There were theories galore. Systems. Functions. Roles. There was no passion, no joy, no triumph, no envy, no lust, no hate, no greed and avarice, cowardice, or dreams. Like the little old lady in the Wendy's commercial, I cried, "Where's the beef?" This stuff did not in any way conform to what

I had seen and lived "out there." It seemed to reduce leadership to a task anyone could be taught and management to a kind of paint-by-numbers art—stay within the lines (of reengineering or of participative management, for example) and you will have a pretty picture. Tell that to Picasso and Van Gogh, I thought. Tell that to Proust, or Dostoevsky, or Mark Twain. Tell it to George Washington, Martin Luther King, Abraham Lincoln, Winston Churchill, and Charles de Gaulle. By extension, tell it to Lee Iacocca, Ted Turner, Stephen Jobs, and Bill Gates. It won't wash. It's not true. Paint-by-numbers became my guiding metaphor as I began a long search for the golden fleece of real art and real artists in management. Like most people, in the beginning I was caught in the established stereotypes: There are "leaders" (the good guys) and there are "managers" (the bad guys).

However, when you take a trip, you see things along the way. The study of art, and the search for artists in management, helped me to see the real *Artists* all right, but it also gave me an intellectual lens through which I saw their fellow travelers, the *Craftsmen.* That discovery would prove critical, as you will see later. In addition, the lens helped me to see the real paint-by-numbers managers, the enemies of both art and craft—the *Technocrats,* those for whom "the technical side of an issue takes precedence over the social and human consequences."[2] I use that word *enemies* deliberately. A lot of people lament the absence of great leaders today. Many of us cast a nostalgic eye on the Churchills, the de Gaulles, the Lincolns. Great leaders have always been, and always will be, rare—the cream that rose to the top before milk and men were homogenized. But, as I hope to show here, they are more numerous than it first appears, and cream is not the only substance that has a tendency to rise to the surface of things. Leaders have powerful enemies. There's a war going on out there, and it is not for the faint of heart. The good guys do not always win.

It's exceedingly important to understand this war. Ever since the "discipline" of management muscled its way into universities, much to the consternation of many scholars past and present (Thorstein Veblen for one; Allan Bloom for another), it has gained more and more respectability. If management is a *science,* then it stands to reason that everywhere there is a need for management there is a need for the science. So much has this become the prevailing view that a recent French prime minister could say with a straight face, "*On ne gère jamais assez*"—We can never manage too much.[3] In North America at least (for reasons we will explore later, I think that Europe and Japan are partly inoculated), we have brought the so-called science of management to government and to social institutions like hospitals and schools. Now, it's true that some disciplines within management—finance, accounting, even marketing—do have more of the character of a formal science, but general management is not one of them. To take what we have allegedly proved in general management and graft it onto the public life is to commit a cardinal sin. To imagine, as some do, that modern management techniques eliminate the need for inspiration, intuition, judgment, and the careful selection of the best people is not just dangerous for corporations, national competitiveness, and economic prosperity, it is very dangerous for our societies as a whole. We need to recover some truth.

Of course there are all kinds of truths—literary truth, spiritual and religious truth, scientific truth, sociological truth—and no one can pretend to capture Truth with a capital *T.* The truth I describe in this book is a small truth, a partial truth, one which emerged as the result of studying, up close, 15 living, breathing, CEOs over the 15-year life-span of a global financial corporation. I had this extraordinarily privileged insider's view of the international corporate world because, as a woman who knew something about the economy, I had been invited to serve on a number of boards of directors (at the time that it became politically correct to have at least one on your board). So, while for reasons of confidentiality I am unable to reveal the names, this story is not fiction and the characters you will meet here are not composites but real people, although, as in *Dragnet,* the names have been changed to protect the innocent. I describe and analyze this cast of characters, people who interacted on an organizational stage, and I describe their interactions as the working out of the different character types which I measured but did not invent: the *Artist,* the *Craftsman,* and the *Technocrat.* These three archetypes (real people diverge, of course, from ideal categories, and you will see the nuances later) are described by their peers, immediate subordinates, and members of their boards of directors as follows.

Artist	**Craftsman**	**Technocrat**
Unpredictable	Well-balanced	Cerebral
Funny	Helpful	Difficult
Imaginative	Honest	Uncompromising
Daring	Sensible	Stiff
Intuitive	Responsible	Intense
Exciting	Trustworthy	Detail-oriented
Emotional	Realistic	Determined
Visionary	Steady	Fastidious
Entrepreneurial	Reasonable	Hardheaded
Inspiring	Predictable	No-nonsense

These are the signposts, the characterological underpinnings of behavior. I found out early on that "management" seemed to wish to do away with character. It seemed to wish to carve people up into their various pieces: the eyes that see, the hand that executes, the head that thinks (conveniently forgetting the heart that feels). Having done so, management went on to develop recipes to address the parts: Teach people how to see better, teach them how to think better, teach them how to behave and to feel, and to be nice. Even teach them how to have vision. Right now in America, literally billions of dollars are being spent every year in the misguided attempt to turn *managers* into *leaders.* Well, the hand, the head, and the heart come in packages, and the package is called *character.* It's unrealistic to think that you can teach a stone-hearted person to be nice and to have vision. At least not quickly, and certainly not in a management-training course. If today we want managers who are open-minded, we have to select those who are; this was once self-evident.

We want, so it is alleged, managers who are visionary and who listen, who create "learning organizations" and who love to work in teams. In my experience, and in my research, the only real barrier to teamwork is the technocratic mentality that we have put in power. For 300 years or so we have been taught that emotion interferes with good decision making, with objectivity. So, we have put in power those men and women who seem to be the most objective, the most coldly calculating, the least emotional—the Technocrats. The trouble is, we were wrong about our assumption: Science is now beginning to prove that, while it is true that emotion can impair judgment, its absence also impairs judgment. So, we have had people in power who are highly intellectual, analytically brilliant, and cerebral, but who have demonstrably bad judgment. No small wonder, then, that we have lost our way. Led by the judgmentally blind, we have stumbled.

So, this book hopes to explode three interrelated myths. The first myth is that we need one kind of leader—the charismatic visionary. The second, and much more pernicious, myth is that anybody can become one. The third is that if you tell leaders what they should do, they will all do it.

Myth 1. Was George Washington a charismatic visionary, or was he "just" a wise and honest man and the leader America needed at that moment? Winston Churchill *was* a charismatic visionary, and he was perhaps the only leader who could have saved Great Britain during the war. He was also moody, autocratic, stubborn, and sometimes dead wrong. Leading is all about the interaction of a leader with timing and context.

Myth 2. If we had sent the prudent man George Washington to leadership seminars or to the Harvard Business School, could he have become a visionary? What about if we had taught him how to think "laterally," would that have worked? Would it have been possible for Winston Churchill to learn to become calm and wise instead of impulsive and dictatorial? Even if we could train people to be what they are not, for the life of me I can't figure out why anyone would *want* to transform a wise leader into a visionary one or vice versa. Leading is all about wise people and visionary people working together.

Myth 3. We claim that we want only leaders who *listen* to their employees. If Winston Churchill and Charles de Gaulle and Franklin Roosevelt had "listened," the Allies would have lost the war. Leaders who listen are not usually the same leaders who inspire. In many ways they inspire because they *don't* listen. They follow their dreams. You probably could have gotten Churchill to admit that he ought to listen to other people, but that doesn't mean he could have done it. You can teach people the vocabulary of leadership, but vocabulary is no guarantee that when the chips are down they'll be able to follow the words with the actions. They can't walk the talk. There is a group of leaders out there—Technocrats—who have all the vocabulary of imagination and listening, but no seminars on earth could make them into visionary people or wise people. They're rigid and dogmatic and cold and calculating, and they don't listen to anybody. *Leading* is about putting them in their place.

This book is about real people, *leaders.* Some who build and some who destroy. If you need a dream, get a dreamer. If you need realism, get a wise man. If you've got an ugly job to be done, get a calculator. If you need all three, build a team and keep each one in the right place. And stop throwing good money after bad trying to transform lead into gold, sows' ears into silk purses.

Part 1, "The Players and the Plot," is descriptive. In Chapter 1 we find the Artist, the "administrative genius." He's imaginative, intuitive, funny, inspiring, exciting, and emotionally volatile. He's visionary. He can be moody, sometimes solitary. People as seemingly diverse as Abraham Lincoln and Winston Churchill and Ted Turner and Walt Disney give some idea of what my Artist might look like. Most of the Artists you will meet in these pages would be dismissed and discounted and run out of town on the grounds that they were "unprofessional," "dreamers," or "fools."

Chapter 2 describes the organizational Craftsman: dedicated, trustworthy, honest, stable, realistic, and wise. We will see that these qualities are very much in disrepute, in part because modernity cannot suffer authority, discipline, and tradition—the craft virtues.

Chapter 3 portrays the Technocrat. Cerebral, stiff, uncompromising, intense, determined, hardheaded, meticulous, often brilliant, he or she pretends to want *reason* to dominate *emotion,* but this is a lie. All sane people want reason to be in the driver's seat; the Technocrat wants, passionately, for reason to crush passion.

Chapter 4 puts all three characters into perspective, in relation to one another. It shows what they think of one another and foreshadows the kinds of conflicts that will emerge.

Part 2, "The Play," begins by describing how these three character types worked together and against one another over the 15-year life of a multi-billion-dollar, multinational financial organization. It shows how one bad succession decision (small business beware!) led to a downward spiral of events that saw the Technocrats triumph and caused the ultimate demise of the organization. It was killed in cold blood, or, more precisely, it died because of cold-blooded "objectivity" and "professional" management.

In Chapter 6 I try to explain the reasons for the technocratic victory and spell out its organizational consequences. I argue that the Technocrat manages to hijack the organization with our support and complicity, and that the pseudoscience of futurism must now bear part of the blame.

Part 3, "The Moral of the Story," spells out some of the consequences for the rest of us. Chapter 9 revisits current theories of leadership, argues that they have become the cure that is worse than the disease, and ends with a description of the nine types that emerged in my research. Chapter 10, "Where There's Smoke There's Apt to Be Fire," is destined to help the reader see these types more clearly in other organizations, like CNN, Marriott, and Wal-Mart. It shows that there are some great American institutions—for example, 3M—still firmly (thank Heavens) in the hands of Craft leadership. Chapter 11, "About Teams," disputes current naive notions of building teamwork and argues that the only barrier to teamwork is Technocrats. For obvious reasons, then, Chapter 12, "The Scope of the Problem," is forced to speculate about the numbers of Technocrats in power in the private (large and small business) and public sectors. A final chapter, "Partial Solutions," offers a diagnosis of the origins of the technocratic triumph and, as the title implies, some hesitant solutions.

My conclusion is brief, because I don't have any magic potions to offer. So, what you will not find here are any new formulae for managing: no appeals to "flat structures," no "reengineering," no "participative management," no instant just-add-water leadership recipes. You *will* find description and analysis. Management scholars have been too hasty in their prescriptions; it's all very well and good to tell managers what they should be doing, but what if the people out there are simply incapable of following the recipe? This book is deliberately and consciously descriptive because description is the best ally of healthy change. Accurate, compelling description can change the world. Accurate description can help people to better understand their world, to recognize that it is they, not us, who are best placed to judge what to do with it.

I've used more imagery and metaphors than technical language, because technical language, though it seems impressive, often does little to enlighten. Pseudoscientific language is not more objective than poetic language, and "we delude ourselves if we think that philosophical or critical language for these matters is somehow more hard-edged and more free from personal index than that of poets or novelists."[4] Further, the tone of this book is passionate. I believe that passion, caring about something, increases our ability to see clearly. If we care, we look more *care*fully. We are more, not less, scientific. This, too, is now a sacrilege.

People interested in testing what kind of leader they are can use Appendix A. Those who are interested in intellectual travelogues can learn about why I came to use art as a guidepost in Appendix B, and about my methods, in Appendix C. Everybody else, follow me. Let's have a look at what I found on my trip.

Part 1

The Players and the Plot

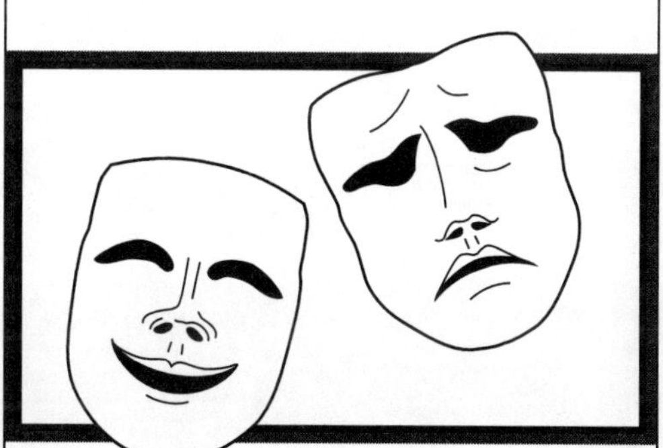

1

The Artist

Ah, but a man's reach must exceed his grasp,
Or, what's a heaven for?

Robert Browning, *Andrea del Sarto*

He walks into the room and everyone smiles. Everyone, that is, except the Technocrats. They remain impassive. Sometimes, slight scowls of irritation cross their faces to disappear as rapidly as they came. They resume their serious conversations. Slowly but surely, the others all move toward him, to shake hands, to bask in that broad smile, to be around him, to be affected by his engaging energy and charm. He has a personal word for each one. "How was your trip from Hong Kong?" "How's that project you were working on? Is it going well? Can I help?" "How's Mary? Did she get over that flu? Tell her Sally sends her best. We'll be in New York next month and we'll get to see you both." Someone whispers a joke in his ear; he laughs uproariously. Then, almost reluctantly, he says, "Well, I guess we better get down to business."

He is tall, strong, with the flashing eyes that betray the intelligence and vivacity behind them. Around those eyes, fanning out above and below, we see the only lines on his face—laugh lines. His bearing, his silvery gray hair, lend him an air of wisdom and class. Truly a fine example of the male of the species, an aristocrat. But he's not alone. There are two others like him in the room.

One of them is slightly shorter but still very strongly built, tall and broad-chested. He, too, has gray hair. He, too, laughs easily, heartily, and his face bears witness in the form of deep laugh lines running from nose to mouth. But he's more tense, rigid, excitable. He makes me think of a nervous thoroughbred just before the big race. His eyes lack the softness of the first; they are more penetrating, intense. His forehead is deeply creased; there are telltale furrows of stress and fatigue and, maybe, of a hard night before. His greetings are warm—a grin, a brief hug—but shorter in duration, and he has a slightly distracted manner, as though he has a dozen other things on his mind. And he does.

The third one is suave. I know this word has fallen out of use in an age that sees Old World manners in a man as sexist. But *suave* is the word I want. He's smooth, not in the modern pejorative sense of ingratiating, but smooth in the sense of soft, not abrasive. His smile is genuine and warm. His voice is pleasantly modulated, his chuckle vaguely seductive. He wants every woman in the room to feel like a queen, especially the older and/or not-so-attractive among them. Some of these women, the brittle ones, bristle on contact. The others primp and preen, feel special. His manner is not intended to manipulate or to condescend. He was raised to be polite and, in his world and in his day, part of being polite was to make a "lady" feel good about herself. This smooth exterior belies the force, the drive, the energy, the determination that lurk below the surface. I'm sure he has often been fatally underestimated by his foes.

These three men are very different. Each was born in a different world, at a different time and a different place. Their family constellations were different: One lost his father very young, one had a happy childhood, one was more troubled. Their educational backgrounds are poles apart. One has a lot of formal education, degrees in both law and business, and two were apprenticed to their trade. One worked all his professional life in the same business; one, half his life; the other has had many different jobs. At lunch with them in January 1994, two and a half years after the completion of my study, I asked them if it felt odd to be lumped into the same sack. "No, not at all," they said. When you respect someone, you don't mind the comparison.

For, in spite of their vast differences, these men are the same. They are leaders, visionary leaders—in my jargon, *Artists.* Let's call them, respectively, James, Cobb, and Mike. We are in a board meeting with these three men. James is CEO of the worldwide group, and Mike and Cobb are CEOs of the Canadian and European holdings, respectively. Each of them will play a starring role in the drama that is about to unfold, a drama which will see their demise and which will reduce the corporation from greatness to ashes. As I explain in Appendix A, I asked 10 of their colleagues—friends and enemies—to give me a reading of these 3 men and 12 others. All those responses were fed into a computer program that gave me a mathematically generated map of all the characters.

First, let's zoom in on one part of that map. Later, we'll look at the whole picture. As you can see from the chart on page 16, James, Mike, and Cobb are surrounded by a cloud of adjectives. At the center are the core adjectives—*inspiring, funny, entrepreneurial, visionary, intuitive, emotional.* Radiating outwards, we find *imaginative, bold, daring, unpredictable, volatile, exciting, people-oriented, easygoing, warm, generous.* Barely outside of the outer circle, we pick up *open-minded,* a quality they share with the Craftsmen.

Without getting into the mathematics, let me try to explain what this means. Just because the word *inspiring* sits next to Mike does not mean that Mike is the most inspiring of the three. The position of any person or any adjective is determined jointly by the mathematical impact of every other adjective or person. In the computer algorithm, these particular adjectives and people were fighting other adjectives and people; for example, *open-minded* was mathematically drawn to both the Artist and the Craftsman and therefore ended up in the core area of neither. The

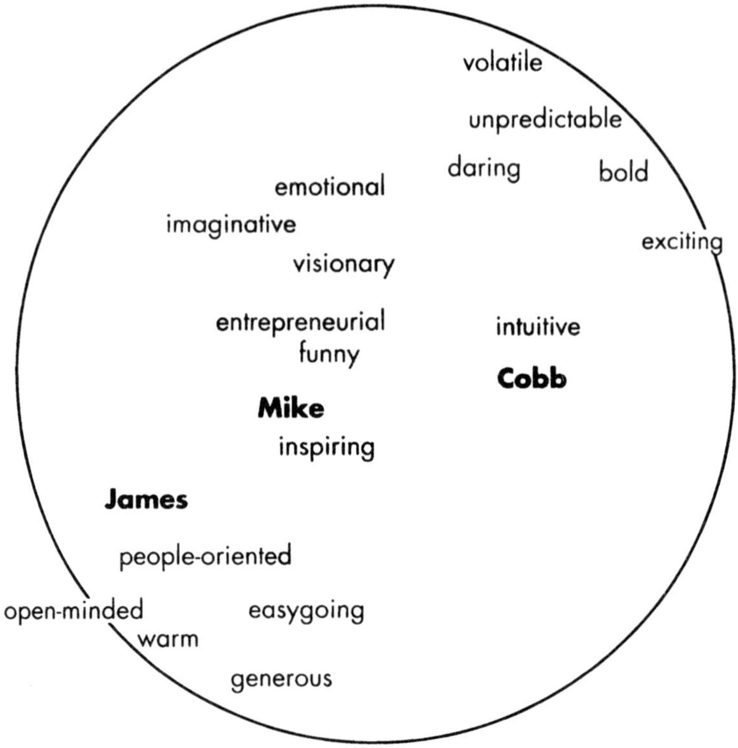

cluster of adjectives we see here represents what I call the *Artist Ideal Type.* Real people diverge from this Ideal Type by degrees. The adjective *exciting,* for example, is highly correlated with the adjective *bold* and more weakly correlated with the adjective *warm,* but James is exciting and warm. It's all a matter of degrees. The farther out you get from the core, the less the person conforms to the archetype. James, for instance, shares many qualities of the Craftsman, which we will explore in the next chapter. Collectively, these adjectives are signposts or, if you will, the fingerprint of the visionary leader. Remember, they were produced by people who liked them and people who despised them.

There are very different kinds of qualities going on here. Let's take them in groups: behavior, thought processes, and temperament.

Behavior

This group might loosely be called *action,* or *behavior,* words: *entrepreneurial, bold, daring, unpredictable, exciting.* Unpredictable, because the design is only in their heads—not on paper. An occasion comes along to complete a piece of the design so they seize it: Carpe diem! They're entrepreneurial. To the outside world, this may look to be, at best, opportunism and, at worst, madness, but in all cases it is unpredictable. It will also seem bold and daring because it is very apt to defy conventional wisdom of the day: A camera does not take instant pictures (Polaroid); a computer is never friendly (Macintosh); *family* doesn't go with *sports car* (Mustang). For Pete's sake, these things just aren't done! In this world, the world of James and Mike and Cobb, the equivalent was "Banks simply do not buy cable companies." But they don't like to bow to conventional wisdom. So, they create excitement.

Thought Processes

Then there are what you might call *qualities of the mind* or *thought processes: imaginative, visionary, intuitive, open-minded.* One of James's longtime associates, a woman, said of him:

> I was always astonished and puzzled by his capacity to sense the waves of the future before anyone else. Sometimes I thought of it as feminine intuition; the French call it *pif,* the nose for it. Like some kind of sixth sense.

This is common to all the responses to these men. They think very long-term. Cobb talks about strategy:

> To develop a strategy you have to have a vision. I have a vision for twenty years, but then I have to bring my vision down to five years

> and to three . . . You work backwards to strategy. It can't be born out of day-to-day perception.

Is it based on gathering all the right facts? I asked him.

> The strategy comes from astrology, quirks, dreams, love affairs, science fiction, perception of society, some madness probably, ability to guess. It's clear but fluid. Action brings precision. Very vague but becomes clear in the act of transformation. Creation is the storm.

Doesn't this remind you of the poet Spender, who tells us of his own poetic inspiration that there was only "the dim cloud of an idea which I feel must be condensed into a shower of words"? Cobb tells us that this vague, indefinable, long-term vision is clarified only in action and, he later adds, by remaining open to new insights.

Why, then, do people follow these guys if they have no idea where they're going? I would argue that they do know where they're going, but it's vague, more a trip than a destination. Visionary leaders inspire with metaphors rather than with detailed descriptions of the future. James talked repeatedly, in the early days, of a branching tree. According to his contemporaries, these metaphors had to be decoded.

Since, we are always told, a picture is worth a thousand words, let me economize a little on the words. I claim that the visions of the visionary leader are no different in form or origin that those of an artist. If you ask a great painter what he or she's going to paint next, it's a rare one who will have a detailed answer and if he or she does, I doubt he or she satisfies the definition of *great.* Studies show that better artists stay radically open as they work on a canvas; there is a continuous interaction between a vague vision and the concrete act of painting. Take the following reproduction of a painting by Salvador Dali.

You can't tell me that Dali had a clear mental picture of this painting before he painted it. All of aesthetics tells us otherwise. Now look at the graphic following it. This is my attempt to capture

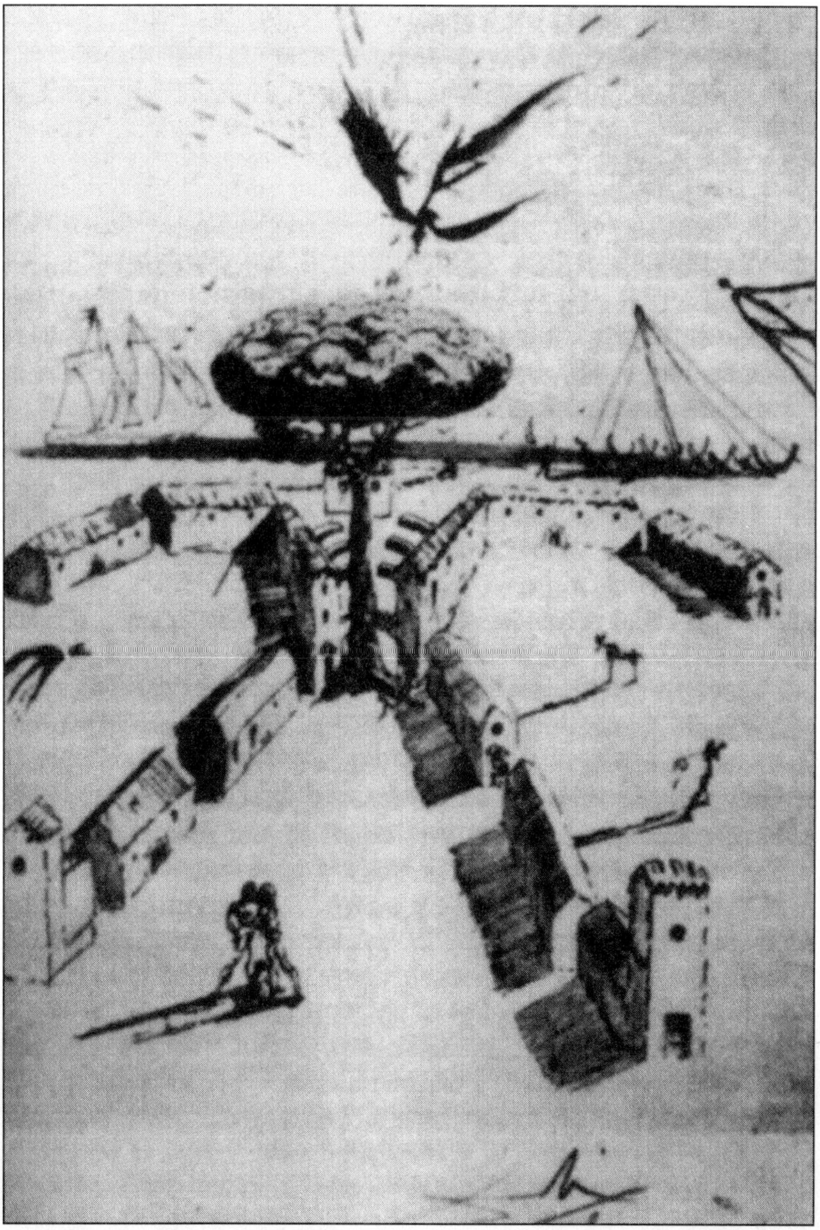

The Beach of Sète by Salvador Dali.

James's metaphor of the branching tree, which for him way back in the 1960s, sitting in a small regional capital of the Western world, meant something about a major international, integrated, financial-services company at least 20 years before such a strategy became a modern, conventional notion.

James had a design, but it was inchoate, unclear, fragmentary, "nothing but a germ ab initio," and only in his head. How could it be otherwise? Conceptually, it didn't exist, and at the time the conventional wisdom was "Bankers are bankers and insurers are insurers, and never the twain shall meet": the so-called four

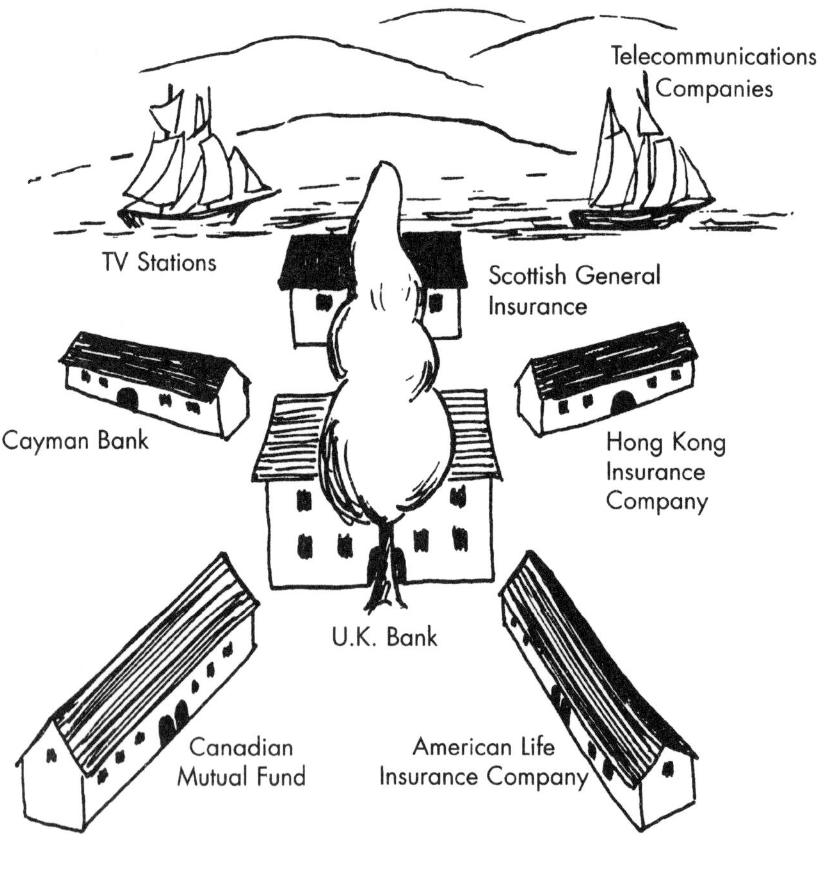

pillars of the financial world (banking, trust, insurance, and brokerage) were each inviolable. To complete his "painting," he needed time and power. Artists are notoriously jealous of their artistic control. Imagine someone having the power to say to Dali, halfway through the work, "That silly-looking tree in the center doesn't make sense; take it out." Each piece of the dream that James put together—from a U.K. bank to a Canadian mutual fund to cable companies on the horizon, ostensibly unrelated to financial services—was, for him, indispensable to the design, although he could not have articulated it beforehand. So he talked in metaphors and seized occasions and was accused of opportunism, of working without a plan.

Artistic vision always seems to have this fragmentary quality. Beethoven is said to have composed the whole of the Hammerclavier Sonata beginning with a point of rupture between two broad movements that had yet to be conceived.[1] In a letter to his mother, Mark Twain describes such a phenomenon:

> I am trying to think out a short story. I've got the closing sentence of it all arranged and it is good and strong, but I haven't got any of the rest of the story yet.[2]

How do you explain to a nonbeliever that you indeed have an idea for a short story, a sonata, a strategic vision for your organization? Discursive logic, language, is inadequate to the task, but a metaphor might do it.

And, to come back to my main theme of why people followed them, now we have a vision and it's wedded to action, fast action, exciting action, death-defying high-wire acts. This is already enough to cause people to sit up and take notice and for some to wish to follow. Coupled with the other group of qualifiers, you have a knockout. These are, let us say, the qualities of the heart, of temperament.

Temperament

People-oriented, easygoing, warm, generous, funny, emotional, and *volatile.* I began this chapter talking about how James came into the room and always, without fail, had a warm personal word for everyone. To be sure, he was also task-oriented and extremely busy, but not so busy as to ignore people. James and Mike and Cobb are all people-oriented, not in the sense of being preoccupied with how to motivate them, but in the sense that they take people, their feelings, their needs, their worries and fears, into account. All is not a calculus. Cobb talks about charisma:

> I always believed that a work environment is fun if it has a drive, it has a sense of purpose, a sense of making people feel that they're important, they're achieving something. So I think it was instilling that. . . . It's a simple message, but it strikes down deep in the heart of people and in their feelings. . . . All of a sudden you have allowed them to bring out their emotions, what they feel deep down which never in business are they allowed to do otherwise. . . . That's probably what charisma's all about. Charisma is not about things. Charisma is something of the soul. And because I was bringing it out, others could bring it out and it was contagious, and this is why people were so excited. They were having fun in what they were doing and the pleasure comes, I think, from the fact that your soul is part of the game as well as your mind.

And James dared to say publicly:

> Man reacts instinctively to the pressures of material need and the desire to live in friendship and love. It seems to us that to respond to these twin needs gives our institution a worthy reason for being.

How many CEOs do you know who write about love in their annual reports? And how many really believe in a moral purpose for their organizations? One of my Artists talked about the insurance business. His words may surprise you:

> I tell my men that they are in the most moral business in the world. Nobody likes to think about death, talk about death. The life-insurance salesman who gets his foot in the door to sell you a life policy is engaged in a moral battle. There are millions of widows and children all over the world who live a better life because some pain-in-the-neck sold a policy to an irritated father.

Add a drop of *generous* and *easygoing* to this curious mixture. What do *generous* and *easygoing* mean? They mean you don't get fired if you make one mistake. Let's listen to Cobb again:

> You have to learn to have confidence in your people, to allow them to run, take some chances on them. I kept saying, we have to accept that we're going to make mistakes.

Finally, I've come to the last two adjectives: *emotional* and *volatile*. I've left them for the last on purpose, because they are going to allow me to tell another part of the story—the story about the inner life of the Artist. I'm going to tread very softly here because I know a lot of people get uncomfortable when someone talks about psychodynamic principles. Why, a few actually run out of the room at the very mention of Freud or Jung! Hold on. Stay with me. It's important.

Before I get deeply into this, I have to clarify a couple of things. First, I will say nothing about any single individual by name here; such information is entirely too personal and, therefore, out of bounds. Second, I'm adding two other people with artistic leanings, Bill and Ross, to the artistic group to further protect anonymity (see Appendix C).

Okay. Do you know about the Minnesota Multiphasic Personality Inventory (MMPI)? It's a personality test. Not one based on 20 or so questions (your garden-variety quick-and-dirty test) but on 566 questions. The results were machine-scored to ensure uniformity and anonymity, and the interpretations were made by a clinical psychologist who did not know the players. The MMPI results, along with my observations of a decade and the results of all my interviews, form the basis of the next section.

Inner Life

Man was made for Joy and Woe;
And when this we rightly know
Thro' the World we safely go.

William Blake, *Auguries of Innocence*

Emotional. Volatile. Joy and woe. Up and down in both directions. Like Lincoln, Churchill, Dickens, Cole Porter, and Monet. Churchill called his depressions "the black dog"; they haunted him, stalked him. Charles Dickens tells us, "I am sick, giddy, and capriciously despondent . . . I have bad nights, am full of disquietude and anxiety . . . my wretchedness just now is inconceivable." And yet, his prevailing mood, the dominant one, was one of elation and energy and prolific inventiveness. So, for the most part, it's joy and excitement these Artists convey, openly and without apology.

To set the stage a little for what follows, let's take a closer look at Winston Churchill. Recognized as visionary even by his enemies, Churchill, as a military strategist, acted on intuition. (Of course, since intuition can be as radically wrong as right, he had major failures as well as successes.) He was eloquent and funny and regularly used metaphors to convey his ideas; it is to him we owe the post–World War II expression *iron curtain.* His sentences were not marked by rigor or by logic but by metaphor and passion. Before World War II, he mercilessly harangued the British public, warning of the dire consequences of a lack of military preparedness. Listen to the metaphor:

> I have watched this famous island descending incontinently, fecklessly, the stairway which leads to a dark gulf. It is a fine broad stair-

> way at the beginning, but after a bit the carpet ends. A little farther on there are only flagstones, and a little farther still these break beneath your feet.

How much more powerful, more mobilizing, is this image than the comparable "logical" expression: "We are devoting only 2.1 percent of the GNP to preparedness while our potential adversaries have, since 1934, devoted 2.13 and 2.16." Who cares?

Churchill's thirst for adventure, his excitement, even jubilation, was the normal face of the man, but there was also a somber side, the depressions that typify the manic-depressive. While manic-depression, or, in medical language, *bipolar affective disorder,*[3] is a serious illness of genetic and physiological base involving hallucinatory episodes, flights of ideas, and complete loss of contact with reality, its milder form—cyclothymia—was certainly evident in Churchill.[4]

No one is clear about the relationship between depression and creativity, but some argue that all the evidence indicates that the deep experience embodied in a classical work of art, for example, must be what is clinically called a *depression* and that the stimulus to create such a perfect whole must lie in the drive to overcome an unusually strong depression.[5] The temporary loss of interest in the external world characteristic of depression may provoke intense activity in the inner world.

Apart from its genetic and physiological base, what might provoke the depression? Arieti, a psychiatrist who has made a serious study of the thought processes characteristic of Artists, speculates that the creative person senses a defect, or incompleteness, in the usual ways of looking at the world and finds them suffocating. He or she needs to break with conventional wisdom. After all, to knock your head repeatedly against the brick wall of complacency is depressing. I asked one of my Artists, "If you could do it all over again, what would you change?" Listen to what he said:

> If I could rewrite my life? Then, the shackles of my youth, I would smash them. I would really destroy them to allow myself to explore my life.

Conventional limits—what to think, what school to go to, how to behave—all these were unbearable to him. I asked what kind of people made him uneasy. He replied instantly:

> Ambassadors [uproarious laughter]. Because they're phonies. . . . And unctuous people. And the cocksure who think they're so much superior to others. The snobs—the snobbier they are the emptier are their heads. The people who have nothing to say. I don't mean the people who are silent. It's people who are so shallow, empty, vain.

In short, the people who mouth clichés and pretend to have ready answers. Conventional wisdoms of all sorts. Pierre Péladeau, a Quebecer who started with a small community newspaper in Montreal and now controls an international printing empire with assets of $3 billion (including America's second-largest printer and France's Fécomme), is manic-depressive, takes lithium to control it, and has never bowed to conventional wisdom in his life. I'm told that CNN's Ted Turner is similar.

All right. There is, for whatever reason—genetic, biological, developmental, or circumstantial—a connection between depression and creativity, but the dominant mood is *enthusiasm.* James wrote:

> Our era demands much but promises even more! Even though we may feel a certain vertigo faced with all the dangers, we can also taste—even exalt in—the joys of knowing, creating, living.

A zest for life in spite of its difficulties and dilemmas—maybe even because of them. All these Artists evince and transmit enthusiasm, but what about this depression business? Let's take a closer look.

The word *depression* is encrusted with all kinds of connotations and superstitions. While some people escape it entirely—the hypermanic—depression is normal. Almost all of us have bad days, sad days, days when we just don't want to do anything. This is mild depression. But depression can wear many faces. It can show itself as a certain lassitude or fatigue or withdrawal, instead of sadness. It can show itself as giddiness, a so-called *smiling depression.* Or as an alcohol dependency. It can show itself as hyperactivity; the person is unable to rest or even to sit down for fear that, in doing so, the "black dog" will attack. The artistic personality, the *cyclothymic,* simply has an intensification of these normal cycles. Whereas when I'm depressed I might go out and buy a new pair of shoes, a manic might go out and buy the Empire State Building and charge it on Visa. A cyclothymic might come up with all sorts of schemes.

All five of my Artists show up hyperthymic, manic, and/or cyclothymic. One of them told me:

> Oh, sure. I've had bouts of depression all my life. Serious depression. But then I just went out and bought another company. It worked every time.

They have rich inner fantasy lives, which they allow themselves to experience. Listen to the blind, clinical interpretation of the MMPI results for one of them:

> Healthy ego. Highly artistic, creative. At times has a hard time distinguishing reality from fantasy. Uses fantasy to overcome anxiety. Exceptionally sensitive, nontraditional male. Quite isolated but has enough energy to interact socially. Quite hysterical. Exceptionally creative and fanciful. Similar to Picasso. Totally nontraditional.

Several important things are going on here. First of all, "healthy ego": Does this mean narcissistic? Not at all. It means strong character, a character capable of living with ambiguity, ambivalence, and doubt, neither succumbing to dogmatism nor flying off in an unreal world. According to their psychological profiles based on the MMPI, artists are typically and paradoxically both sicker and healthier than ordinary mortals. They are sicker in the sense of being bombarded by ideas and fantasies; healthier in the sense of being able to control the process and shape something new and useful out of it. The way Edwin Land (of Polaroid fame) describes his process of invention is a good illustration:

> I find it very important to work intensively for long hours when I am beginning to see solutions to a problem. At such times atavistic competences seem to come welling up. You are handling so many variables at a barely conscious level that you can't afford to be interrupted. If you are, it may take a year to cover the same ground you could otherwise cover in sixty hours.[6]

Second, returning to our group, "At times has a hard time distinguishing reality from fantasy. Uses fantasy to overcome anxiety." Does this mean looney tunes? Sort of. But not really. Have you ever noticed that some people dream in technicolor and some in black and white? That some people don't ever seem to have dreams, awake or asleep? That some people literally hate dreamers? A rich fantasy life is not synonymous with madness, though it may be the kind of madness Plato associated with the poet. We have to be able to conjure up a new world if we are then to create it. Or, as Santayana tells us, "Man's progress has a poetic phase in which he imagines the world and a scientific phase in which he sifts and tests what he imagines."

Two of these men showed significant degrees of autism in their MMPI profiles. Fantasy. Living in a world of one's own. Thus, the frequent solitariness of these men. But if the world is too much for them, if their days are always absorbed by pressing concerns, when can they think? They need, periodically, to get away from the noise and the bustle, if only in their heads.

They are all nonconformists. But not nonconformists in the superficial sense of dress—they dress conservatively—or in the causes they espouse. They are nonconformist in the deeper sense, in the sense of skepticism of all conventional and popular ways of thinking, modern or ancient. They take almost nothing for granted. They are, therefore, radically open to new thought, to new associations and combinations, which some feel to be the sine qua non of creativity, like, again, *sports car* and *family,* or *instant* and *photography.*

To summarize, my Artists are emotional and sometimes moody, volatile and funny. They are nonconformist and frustrated by gospel with a small *g.* These characteristics, features of their basic psychological functioning, mean that to the outside world they appear imaginative, intuitive, and visionary. Behaviorally, they are daring, adventurous, and entrepreneurial, and, therefore, for some, exciting. In addition, they like people, so they are *inspiring.* Feeling or temperament, thought processes, and behavior come, as I tried to emphasize earlier, in packages. The package is character—in this case, the Artist character.

The Artist Character

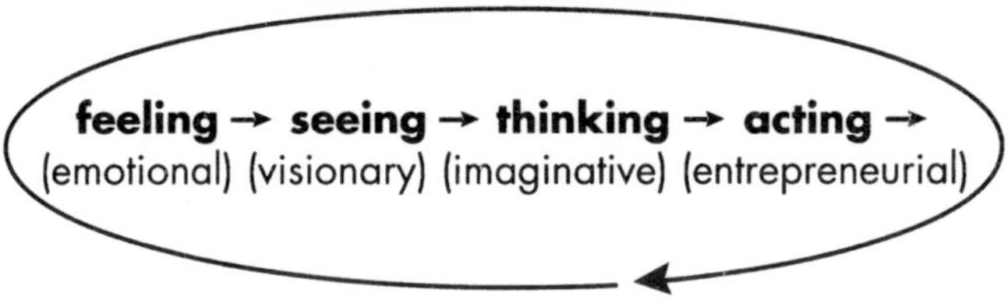

How could we have imagined that someone who is not moody and volatile and emotional could be taught to be intuitive and visionary and entrepreneurial? It's a mystery to me.

You may have already guessed what the Artist's Achilles' heel is, but you'll have to wait until Part 2. Because now we are going to turn to his fellow traveler, the Craftsman, with whom he has a mutual admiration society.

2

The Craftsman

His mind was great and powerful, without being of the very first order; his penetration was strong, though not so accurate as that of a Newton, Bacon or Locke; and, as far as he saw, no judgment was ever sounder. Perhaps the strongest feature in his character was prudence. . . . His integrity was most pure, his justice the most inflexible I have ever known. . . .

Thomas Jefferson's description of George Washington[1]

Calm normally, unless inflamed by injustice or obtuseness. Steady. Solid. Reasonable. Wise. Prudent. Measures his words. Illustrates his principles with real-life stories. Listens. Weighs. Pauses. Having a conversation with him is a pleasure, a genuine, honest-to-God pleasure. He does not begin with "Okay, what do you want to know?" He doesn't presume to know anything perfectly and can't imagine why I want to talk with him. He's not a theoretical man, and because he thinks I'm a theoretical woman, he thinks he'll have nothing to say. In fact, he has a lot to say, but you have to learn to read between the lines as he recounts seemingly mundane events: "So I told him that he'd make a great consultant but that, as a manager, he was a complete washout." Pretty run-of-the-mill stuff, no? Trouble is, the man he was talking about was his boss. So was the man he was talking *to*.

When I first started thinking about the Craftsman, my guiding metaphor was a carpenter. I like wood. It feels nice. There's this man who built our house. His name is Gaétan Roy, and he's a carpenter. He's had a key to our house for the past eight years (he's very trustworthy). He whistles while he works, even on Sunday at midnight because he promised (he's very reliable) the job would be done. He likes what he does and he's done it all his life.

> "Gaétan, can we use pine on the floors?"
> "You won't like it. It's too soft. I can get a better deal on spruce and it will stand up better."

He's direct. He's economical.

> "Gaétan, can you have those logs cut to make a facing on that wall?"
> (Skeptical frown.) "I don't think so, but I'll ask the guys at the mill."

He doesn't pretend to know everything. He is not embarrassed by this.

> "What do I owe you?"
> "I logged forty-two hours."
> "Here's your money."

He's honest. I don't have to check up on him.

> "Gaétan, I love what you did in the staircase."
> "It took me a while to figure it out."

He's modest, humble about his inventiveness but proud of his product.

In general, craftsmanship is about three things: pride, skill, and quality. I can't say it better than philosopher Harold Osborne who wrote, in the British Journal of Aesthetics, that craftsmanship

> involves a genuine pride in the process of production itself, a pride which drives a man to make whatever he makes as well as they can

> be made, even beyond the economic considerations of reward. This impulse, which lies at the roots of fine craftsmanship, is now recognized by anthropologists to have existed from the earliest stages of human activity. . . . It is this impulse, this cult of excellence, which through the centuries of prehistory and history led to the perpetuation of traditions of craftsmanship, the rich storehouses of know-how and skill.

There's a lot going on here. "Pride in the process itself": Isn't this a lot like the old saying, "If it's worth doing, it's worth doing well"? This is an instinct that is basic to the Craftsman. He or she doesn't require college courses on the subject. "The rich storehouses of know-how and skill": People, Craftsmen in particular, are the rich storehouses, the carriers, of skill and know-how, savoir-faire. What is skill anyway and how do you get it? The dictionary defines skill as "practiced ability"; skillful as "adroit"; skilled as "not untrained or amateur." *Adroit,* of course, means having dexterity, with its heavy connotations of physical, manual operations. All these denotations and connotations have to do with experience—accumulated, hands-on experience. The Craftsman is proud to use these accumulated skills to produce a high-quality product. He or she is not an amateur.

Craftsmanship has also to do with the difference between using a tool and using a machine. With a tool, a person controls production from start to finish. There is no radical separation between thinking and doing, planning and executing. Unlike a machine, a tool leaves small irregularities and imperfections that cause each product to differ slightly from every other. Each wood—pine, cherry, walnut, mahogany—reacts differently to the tool. Over many years you acquire powers of differentiation and the skills that match them. Osborne continues:

> In general terms, both tools and machines are devices for facilitating man's constant business of adapting his environment to his needs and improving his conditions of life. We speak of "tools" when we mean devices for extending the powers of the organism for manipu-

> lating its environment. A part of craftsmanship consists in a man's understanding the tools of his trade, possessing the skill and dexterity to use them to best advantage. . . . A good craftsman, it is said, does not blame his tools. . . . It is a sign of good craftsmanship to know the right tools for the job and to know how to use them. This distinguishes the true craftsman from the amateur.

I like to think of managerial ideas such as total quality management or reengineering as the tools, and employees as the wood (give my metaphor a chance—I don't mean that employees are blockheads). Each circumstance and each employee reacts differently to the tool, so you need a different tool for each. There is no universal tool for all people and all circumstances. Tools are not machines. Craftsmen differentiate, judge, select the appropriate one according to the wood and the time and the goal. If things go wrong, they don't blame the tool or the wood; they blame themselves and their own faulty judgment. You do not become a Craftsman overnight. And you cannot just read up on it. You have to do it. Now, let's look at the organizational craftsmen.

There are six of them. They all look different. Like the Artists, they are different ages and have different educational backgrounds—accountants, MBAs, historians, apprentices, and one Ph.D.—the best schools in the world or none at all. They come from different cultures: British, American, Canadian, French, and Scot. They have slight temperamental differences, but in some fundamental sense, they are the same.

Look at the following map. Here's how they are seen. In the core, we find, literally piled upon one another, *stable, realistic, reasonable, trustworthy, sensible, polite, responsible, thoughtful, predictable, steady,* and *well-balanced.* To the west, *conventional;* moving northward, *controlled* and *conservative, punctual, hardworking, knowledgeable;* turning east and then south, *straightforward, dedicated, honest, wise, amiable, helpful, humane,* and *open-minded.* Who wouldn't want such a profile? Can you imagine one of these guys rushing out one day to buy

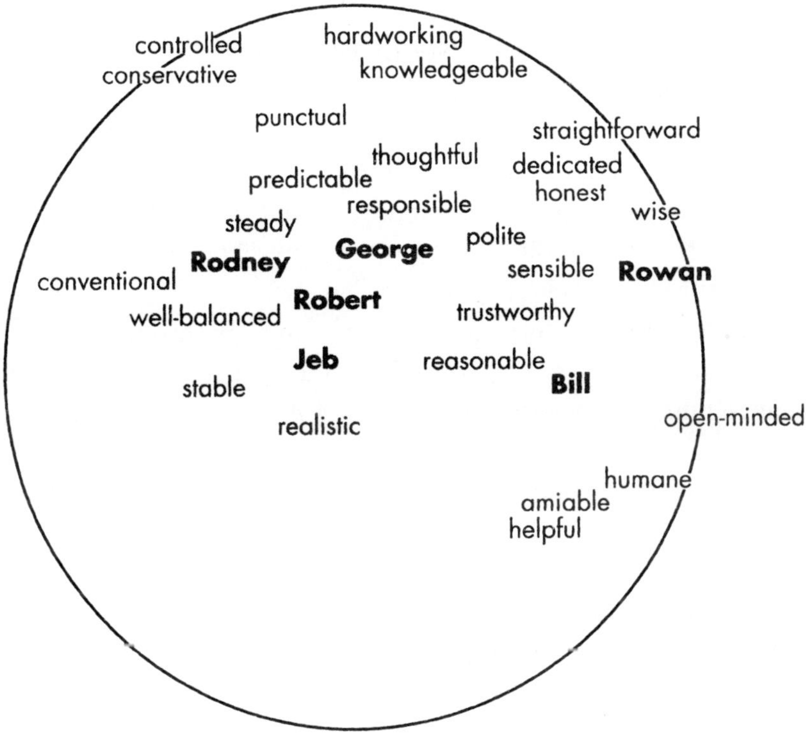

a Corvette and leaving his wife and kids to hitch up with some cute young thing? Not on your life. Can you imagine them experiencing a sudden conversion to anything? How do you think they feel about managerial fads?

To get the flavor of these people, we'll go through the same exercise we performed for the Artists; that is, we'll group those adjectives that seem to speak to different issues—behavior, thought processes, and temperament.

Behavior

The adjectives are *hardworking, dedicated, predictable, helpful, humane, polite,* and *punctual.* They work hard; long hours if there's something urgent, regular hours (say, 10-hour days) if there's not. They'll do whatever they themselves judge necessary for the institution—they're dedicated to its success—but it will be their decision, their judgment. Not a rule. They don't stay at the desk on Saturday just to impress someone with their diligence. Their behavior is predictable; you can count on them to be where they say they'll be, doing what they say they'll be doing. They seldom surprise anyone. They're never erratic.

To accomplish their work, they have to work through and with others, and so they're helpful to them. Like master to journeyman, they're humane but firm: "No, that's not how it's done. Try again." They don't have a screaming fit if someone makes a mistake. They expect mistakes; one told me, "It takes time to weld a team together and to solve human problems. They're not machines you can just turn on and off." Because they respect others, they are polite and punctual. They wouldn't dream of keeping people waiting for them, because they don't want to send the message that they are more important than others are. They are not narcissistic. Let's go to the next level.

Thought Processes

The adjectives are *knowledgeable, thoughtful, realistic, sensible, wise, open-minded, conventional,* and *conservative. Knowledgeable.* Five out of six of these people (one is a very recent recruit) have worked in the same industry all their lives. This is unsurprising. Osborne tells us, "The old craftsman possessed a hard-won skill derived from lifelong immersion in a centuries-old tradition involving inarticulate knowledge of materials and tools, inherited skills and dexterities, principles of design and pride in excellence." When you meet them they say, "I'm an insurer," or, "I'm a banker." They don't say, "I'm in banking," as if to add "right now." They are identified with their traditions. They are proud of them. And they have an intimate knowledge of what those traditions are, how they have evolved and why, though some of this knowledge will be inarticulate tricks of the trade.

Realistic and *sensible.* I put these two together to come up with a third—*good judgment.* Why do we call someone, or something, or some behavior realistic or sensible? Well, let's see. When we say "sensible shoes," we mean *comfortable, useful, practical for walking.* We are trying to draw a distinction between them and, say, shoes that are flashy or for a party. So, there's a connotation of appropriateness to the word *sensible.* Take another instance. "I wanted to go to Tahiti," says a friend, "but given the state of my bank account, I decided on Florida." We respond with, "That was sensible." What do we mean? I think we mean conscious of limits, in this case limits to our debt capacity. Now, we might have said, "That was realistic," but it seems to me that *realistic* carries another connotation. As the root of the word implies, this time we are talking about something conforming to what is real, true, actual, not an illusion. I would like to be a ballerina, but at 46, the idea is no longer realistic; the mind is willing, but the flesh, oh the flesh. . . . If we put all of this together—appropriateness, practicality, limits, and conforming to reality—I think we get *good judgment.*

If we put *good judgment* together with *realistic* and *sensible* and couple it with *knowledgeable,* we get *wise.* The dictionary defines *wisdom* as "experience and knowledge together with sagacious judgment." Young people are rarely wise. Wisdom's one of the few advantages of age, because wisdom takes time and lots and lots of experience—and even then, it's not guaranteed. Time is a necessary but insufficient condition. Some people never get wise, either because they forget the past or because they live only in the future, or both. As Santayana told us in *The Life of Reason,* "Progress, far from consisting in change, depends on retentiveness. Those who cannot remember the past are condemned to repeat it." They are incapable of learning from experience. Or they never stay in any one place or at any one thing long enough to acquire experience. They are gadflies. Or dilettantes. In any case, Craftsmen are wise enough to know that the past has a heavy impact on the present and the future.

Their visions are realistic and, therefore, of medium range, backward and forward in time. They know where they've come from, and where they're going. They see the landscape accurately, and can see to the top of the next hill but not into the valley beyond. Their dreams are anchored in the present and the very near future. If we look again at the Dali painting and we put the Craftsman's product beside it, what would it look like? The first change is that the far distant horizon is no longer visible, although the overall form or configuration of the design remains the same. Craftsmen like to keep the structural integrity of the design. They believe in stability. Ordinary people, they think, need to be able to trust the map in order to orient themselves. Second, there are more people and trees in the design. It is more lifelike, more realistic. Finally, the buildings all have different shapes. The Craftsman is not wedded to uniformity—different tools, different woods.

Then we come to three confusing adjectives: *conservative, conventional,* and *open-minded.* They may seem contradictory, but I want you to think of these three words on a continuum that reads something like "Change if necessary but not necessarily change." All craftsmen work within a tradition, and tradition involves a certain inertia. Where the weight of the past is very strong (in Catholicism, for example), change comes more slowly; the inertia is strong. Where the tradition is newer (rock music, for example), change may be more rapid; the inertia is weaker. Some individuals will be more open to questioning a tradition than others. And it may depend on the situation or the subject—I may be very traditional about men taking out the garbage, but very open-minded about child care. Craftsmen range across this continuum, sometimes conventional in their approach, sometimes all the way to conservative, then back to open-minded. Pushed, they

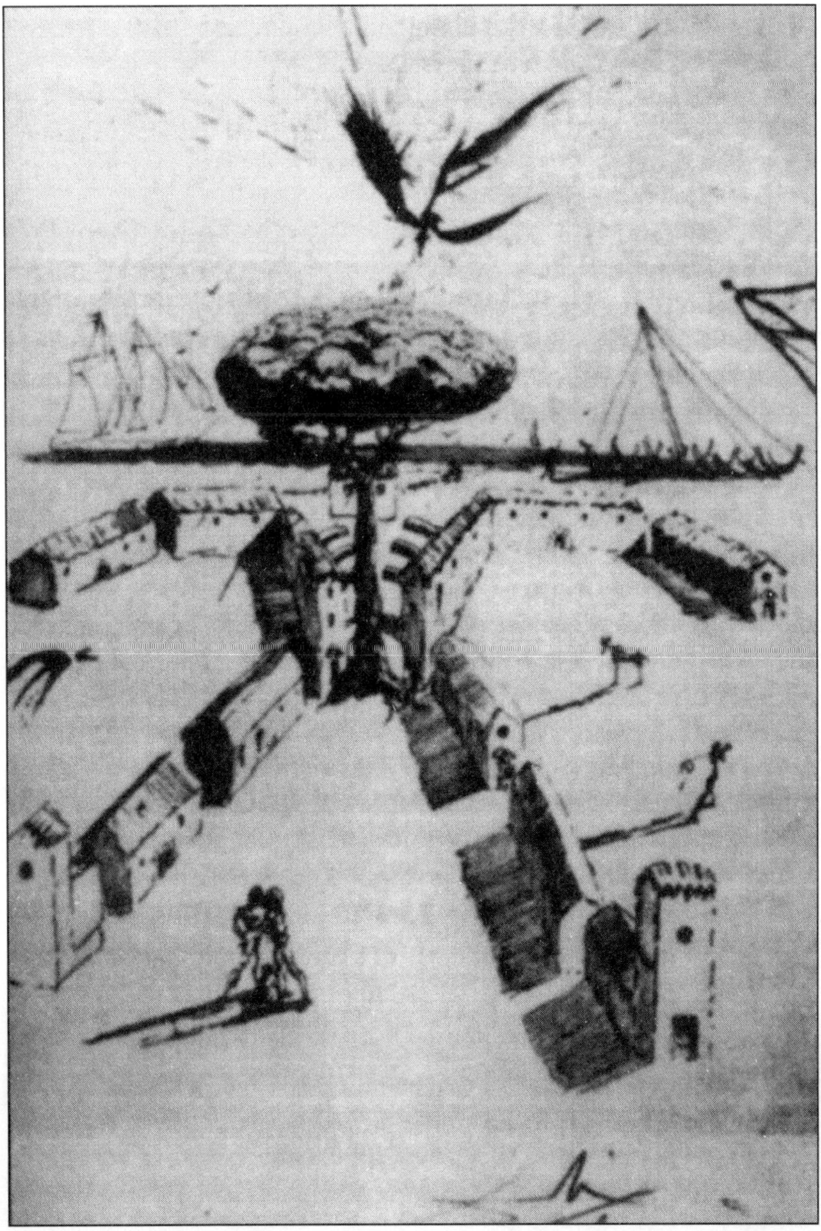

The Beach of Sète by Salvador Dali.

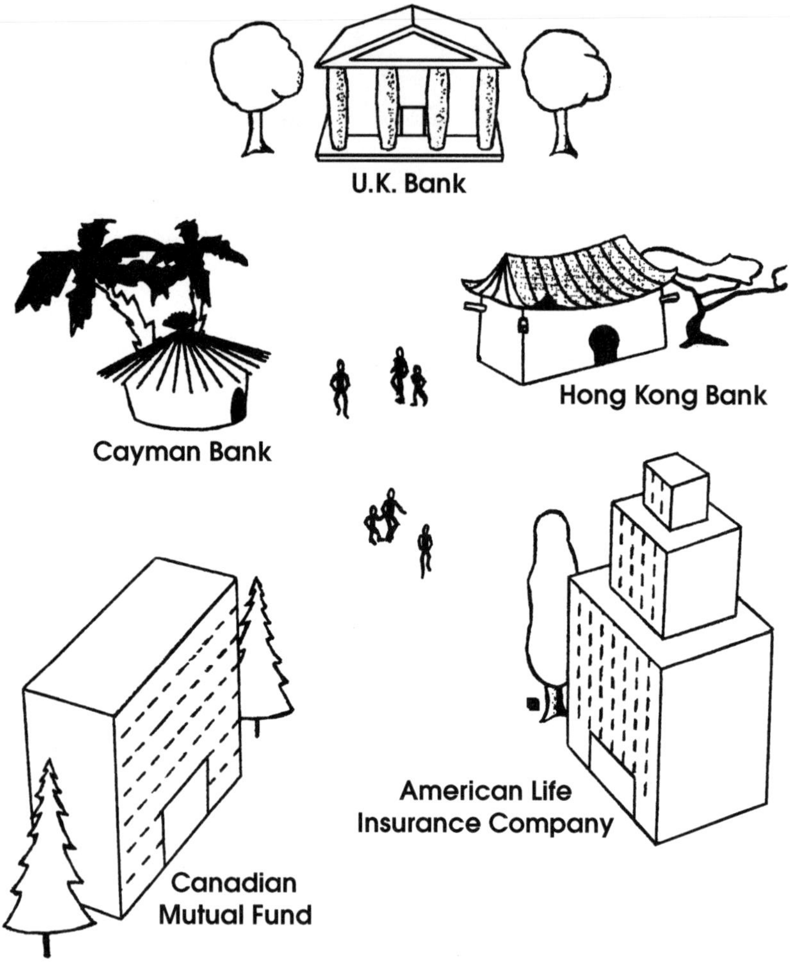

shove, become stubborn, dig in their heels. Coaxed, they listen. Stimulated, they open up. On questions of principle or morality, they can be, like George Washington, downright pigheaded.

Temperament

The adjectives are *straightforward, honest, responsible, amiable, stable, well-balanced, steady,* and *controlled.* Take *straightfor-*

ward and *honest*—Jeb, the Craftsman I mentioned earlier who told his boss he would make a great consultant but was a lousy manager. He didn't say this to be cute, or provocative, or insulting, or insubordinate. He saw it, believed it, and said it—naively maybe, but honestly. No games. No duplicity. Straightforward. What you see is what you get. For the most part amiable, stable, and calm, the Craftsman becomes stiffer, more and more intransigent, if his or her strongly held beliefs are threatened. Like *open-minded* and *conservative, amiable* and *controlled* exist on a continuum, with some Craftsmen in some circumstances more conservative and controlled than they are amiable and open-minded.

All of these qualities add up to *trustworthy.* Craftsmen are reliable and honest, mostly predictable, always sensible, always realistic, and everybody trusts them. They inspire that magical word *loyalty,* because they exhibit it themselves. People are not excited to work for them; they are pleased to work with them. Some people.

Inner Life

It's tempting to describe the Craftsman's inner life as zip, zilch, nada, rien, but that's not true. Only in comparison with the rich fantasy life of the Artist do such words appear to be true. Basically the Craftsman looks like Everyman. Here's a typical interpretation of the MMPI material:

> This must be a Craftsman. Very flat [curve]. Soft, gentle, introspective guy. Just allows the world to happen. No big bitches with the world. Likes people.

This happens to be one of the gentler Craftsmen, but they all share this bit about "no big bitches with the world." They don't have chips on their shoulders. Overall, they like life, find it rewarding. And their mood is usually one, not of buoyancy, but of calm, measured optimism. They expect things to work out in the end. There is actually experimental evidence that so-called normal people wear slightly rose-colored glasses. Apparently we are all born with an optimism thermometer—it's hardwired—and at birth the normals have theirs set ever-so-slightly too high.[2] It's as though someone up there knew that life was going to be pretty tough on us.

"Likes people"; are we talking about some gushing sentimentalist? Not at all. Craftsmen fire people. They fire lazy people because lazy people demoralize the responsible ones. They fire people who are insubordinate; they believe in hierarchy and authority. They are not authoritarian, but the idea that the apprentice should be master to the master craftsman is a contradiction in terms, illogical, unrealistic, silly. They are pitiless with chronic rebels who are more interested in contesting authority than in getting the job done.[3] But they don't like to fire people; they are not sadistic. Nor do they draw some kind of macho pride from it. Bill said to me:

> That S.O.B. [his boss] called me up the other day and his first words were, "How many people have you fired today?" I hung up. It's disgusting.

He found it disgusting because, for the Craftsman, people do not just represent costs; they represent the future capacities of the organization. Bill, in his words, "hides" money in his balance sheet. "It's buried so deep they'll never find it; only I can find it. That's what I use to develop my people." (If the "my" in that sentence resonates with paternalism, that's because it is paternalistic; I, for one, think it's good.) Another one, Rodney, told me:

> People are the organization's most important asset. And the institution was important. People like to feel that they're part of something important, have a sense of belonging, pride.

Neither the institution nor its employees have a purely instrumental value for Craftsmen; they don't use them to attain some other goal, such as personal prestige or power. The institution has intrinsic value.

Craftsmen are not particularly emotional. In fact, the best description might be *reserved;* there are few displays of emotion. They are very far from being sentimentalists, but they do take people into account. They pay attention to them. They invest time in them. And they do like to be challenged. Another Craftsman told me about a former vice president who had been fired. The VP was a difficult guy, but very bright. "He spoke his mind. I'd pay anything to have a guy like that around. He keeps you honest. He didn't agree with you all the time."

Craftsmen know themselves. They know they are not particularly visionary—their vision is medium-term—but that doesn't depress or diminish them. They like themselves. They know their own limits and their own strengths. They are as realistic about themselves as they are about business and life in general. They like the Artists and think every organization needs them. One told me:

> Profit comes from the vision and the people. James and Cobb had the vision. . . . [If you have the vision] and you look after the people, the profit follows. You can't drive at it directly.

The Craftsman works easily with the Artist, chuckles at his or her histrionics, is sometimes frustrated by his or her lack of attention to detail, but "would pay anything to have a guy like that around." This, then, is the craft package, the Craftsman character.

Once again, I ask how could we have imagined that someone who is stable and realistic and wise and responsible could, or should, be taught to be passionate, spontaneous, intuitive, a so-called visionary leader?

I'm going to leave the last words of this chapter to Osborne, because his are better than mine. He's talking about craft in general:

The Craftsman Character

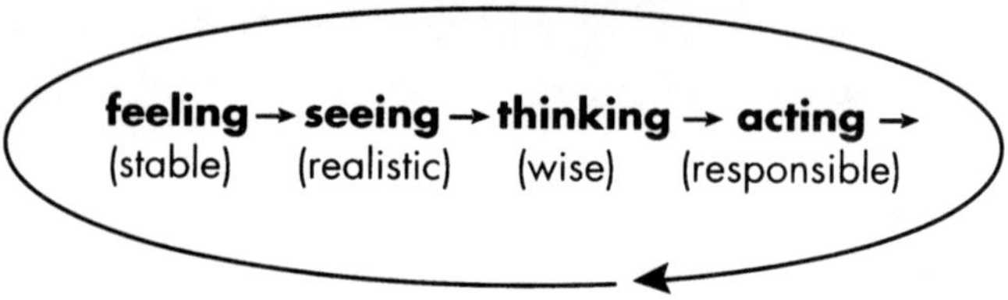

> If the craftsmen who are still active in the world can do something to maintain the cult of excellence, the pride in a good job well done, and the respect for quality, then indeed their contribution to contemporary society is more than justified. They are no *anachronism* but one of the few forces that are left to stem the deterioration and dehumanization of contemporary life. . . . And over the whole field of craftsmanship there is aesthetic delight in the appreciation of skilled and devoted workmanship applied economically to the successful achievement of a practical purpose, while *virtuosity* deployed for its own sake arouses the same sort of distaste as any other form of self-display.

We are now going to meet the virtuoso, who *does* feel that craft is an anachronism.

3

The Technocrat

"Oho, now I know what you are. You are an advocate of Useful Knowledge."

"Certainly."

"Well, allow me to introduce myself to you as an advocate of Ornamental Knowledge. You like the mind to be a neat machine, equipped to work efficiently, if narrowly, and with no extra bits or useless parts. I like the mind to be a dustbin of scraps of brilliant fabric, odd gems, worthless but fascinating curiosities, tinsel, quaint bits of carving, and a reasonable amount of healthy dirt. Shake the machine and it goes out of order; shake the dustbin and it adjusts itself beautifully to its new position."

Robertson Davies, *Tempest-Tost*[1]

The word *technocrat* can be misleading. In my experience, people tend to conjure up a bureaucrat; someone who works (sort of) for the government. But I have in mind a more precise definition: A technocrat is *someone who emphasizes the technical conceptions of a problem to the detriment of their social and human consequences.* Which translates to something like, *I understand that you may lose your house/job/car/child, but my research shows that the system will work better this way and I can't make exceptions.* Now, this person may work quickly or slowly, may be dedicated or lazy, highly intelligent, even brilliant, or just average. What counts is the attachment to studies and to rules, written or unwritten. The unwritten rules are otherwise known as conventional wisdom.

My group of Technocrats is superficially very diverse. Some are slow, even ponderous, in their thinking habits and mannerisms. Some are like hares; they run fast, talk fast, think fast, and argue circles around the competition. They are tall and short and in between. They've had many different jobs in their lifetimes: engineers, consultants, accountants, managers of very different kinds of organizations. Some are exceedingly well-read, others not. They like to collect art—you're supposed to, you know.[2] Some smile and others don't, but when it gets down to business, none of them do. As one said to me, "I laugh as much as the next guy—at home. Work is for working." These people are not funny.

They are rigid. They possess *Truth.* They disdain all those whose reasoning processes are not up to scratch, all those whose logic is infected with, and corrupted by, passion. Colloquially we say, "He has a mind like a steel trap." You can get caught in it like a vise. It's cold, remorseless, even fatal.

Technocrats are difficult to get to know. One Craftsman who worked with them for years told me in frustration, "I don't know what makes them tick. I can't get a grip on them." Personally, I've known some of these Technocrats for more than 10 years, and I still can't get a grip on them, either. You have the sense of talking to a wall—you can try to crash through it, but on the other side you're in just another empty room. It makes you mad, increases your emotional response, and then they dismiss you for the very emotion they themselves have engendered! It's a vicious circle.

Following is the map of how they are seen by their friends and enemies, board members, and peers. At the core: *no-nonsense, intense, determined, uncompromising, fastidious, cerebral,* and *hardheaded.* To the south, where they look a little like Craftsmen, they are recognized as *hardworking.* To the extreme west, *serious, analytical, methodical, detail-oriented,* and *meticu-*

lous. Moving northward, *distant* and *stiff.* In the northeast, *difficult.* Turning southward, *brilliant, insightful,* and *energetic.* You see, when you swing from extreme west to extreme east, you go from ponderous to quick. This is an illustration of the superficial differences that can seem to separate, say, Cam from David and that can mislead us radically. *Brilliance* and *insight,* for example, can and often do masquerade as vision.

Analytical is good. *Methodical* is good. *Serious* can be good. *Brilliant* can be good. *Insightful,* were it coupled with *honest* and *modest,* would certainly be good. Almost any one of these qualifiers, taken individually or in pairs, is complimentary, even flattering. Taken together, we have a syndrome. The Technocrat. Let's look at our groups again: behavior, thought processes, and temperament.

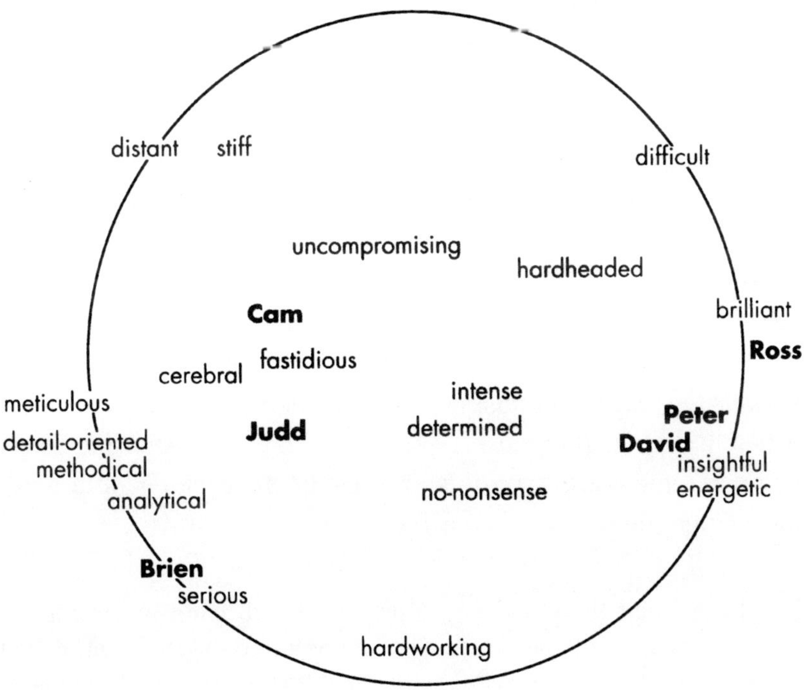

Behavior

He is methodical, fastidious, uncompromising, difficult, determined, energetic, and hardworking. A member of one of the boards had this to say about a Technocrat:

> Very methodical, looked into opportunities very thoroughly. Stubborn. Big ego. Thinks a certain way and so should everyone else. Very arrogant. Well-organized. Brought out his notes from a meeting six months ago and asked what had been done since.

This was no casual observer. He had known this particular Technocrat for 15 years. What he said to me in private, however, he had never dared even to imply in public. He knew from hard experience that he wouldn't be around very long if he didn't keep his thoughts to himself.

Methodical. Fastidious. The *Pocket Oxford Dictionary* defines *fastidious* as "selecting carefully; choosing only what is good." He "looked into opportunities very thoroughly." Yes, but you say, isn't that simply prudent, wise? Then, I reply, how come no one ever calls him wise? No. This was not normal prudence. This was prudence pushed to the edge. This was crossing every *t* and dotting every *i.* This was trying, desperately, to ensure against any error.

Uncompromising. Not uncompromising in the sense of, say, unswerving, uncompromising loyalty in the pursuit of some moral principle, but in the sense of stubborn, big ego. Another board member remembers a particularly distasteful incident in a committee meeting with the same Technocrat:

> He sat right there and we all disagreed with him. Afterwards, he went into the full board meeting and reported exactly what he had argued with us—and lost—as though it was the view of the committee. It was like the committee meeting hadn't even taken place!

I was there, too, and this is exactly what happened. I felt like Alice in Wonderland. He was, to say the least, difficult and hardheaded. You know, to this day, I don't think he did this on purpose. He did it because he was hardheaded in the literal sense of the word *hard,* in the sense of impermeable. It was as though everything we had said could not penetrate. Sort of like water off a duck's back. And he was very determined to have his way.

Yes, the Technocrat works hard and is energetic. But his working hard is not because he's passionately committed to a task, but because working hard is a rule: People are supposed to work hard. It is a measure of the worth of a person how many hours he or she puts in and how few vacations he or she takes. Life is a serious affair.

Thought Processes

The Technocrat is seen as *meticulous, detail-oriented, analytical, insightful,* and *brilliant.* The first three tend rather naturally to go together. If you are meticulous, it means you examine the details of something. If you conduct an analysis of something, it means you look at its parts and their interrelationship; you break down, somewhat artificially, some phenomenon into its constituent parts. Which is what I am doing now—looking at various parts of the Technocrat character to try to understand how it works. But some things do not benefit from being broken down into parts; they lose their essence this way—an idea, a sunset, a poem. The German poet Goethe was furious when people asked him what his novel *Werther* was about; he felt he had struggled long and hard to fuse it into a meaningful whole and people were asking him to break it up again into meaningless parts:

> This made me very angry and I insulted a good many people. For me, in order to answer the question, I would have had to take my little

> book, over which I had brooded so long in order to fuse its various elements into a poetic oneness, and tear it to bits again and destroy its form.[3]

If you are detail-oriented you are likely to be analytical, to want to see the intricate workings of various parts. And when the various parts are all displayed before your eyes, you may be able to pick out some flaws. Then it will be said you are insightful: "Oh yes. I see now. I'd missed that detail. Now I understand. How *insightful* of you to have seen this." In the fifth century B.C., a celebrated Sophist was able to pick out several "flaws" in the first few lines of Homer's *Iliad* and was recognized, by everyone but Socrates, to have been very insightful.[4] But Socrates wasn't big on the Sophists, and Homer's "flawed" poem has lasted quite a while.

In the context of vision, let's look at Dali again, this time with technocratic eyes. The first thing to remember is that the overall design is given to the Technocrat—he doesn't dream it up himself; analysis can only follow creation. In the words of one of them:

> It's not the CEO's job to create the ideas; his job is to sift and measure ideas that come from somewhere or someone else.

Incidentally, this is almost word for word what Santayana told us: "Man's progress has a poetic phase in which he imagines the world and a scientific phase in which he sifts and tests what he imagines." So the Technocrat analyzes the design, breaks it into parts. He discovers that the tree doesn't *belong* in the middle. He discovers that two of the buildings are dilapidated and should be condemned. He discovers that curves—and people—are inefficient and should give way to straight lines and uniform scale. There are no trees here. The drawing is not very "realistic." By the way, we no longer see the horizon. We are too fixed on the details to lift our eyes toward a distant and unknown future; indeed, there is no longer any road opening to the future.

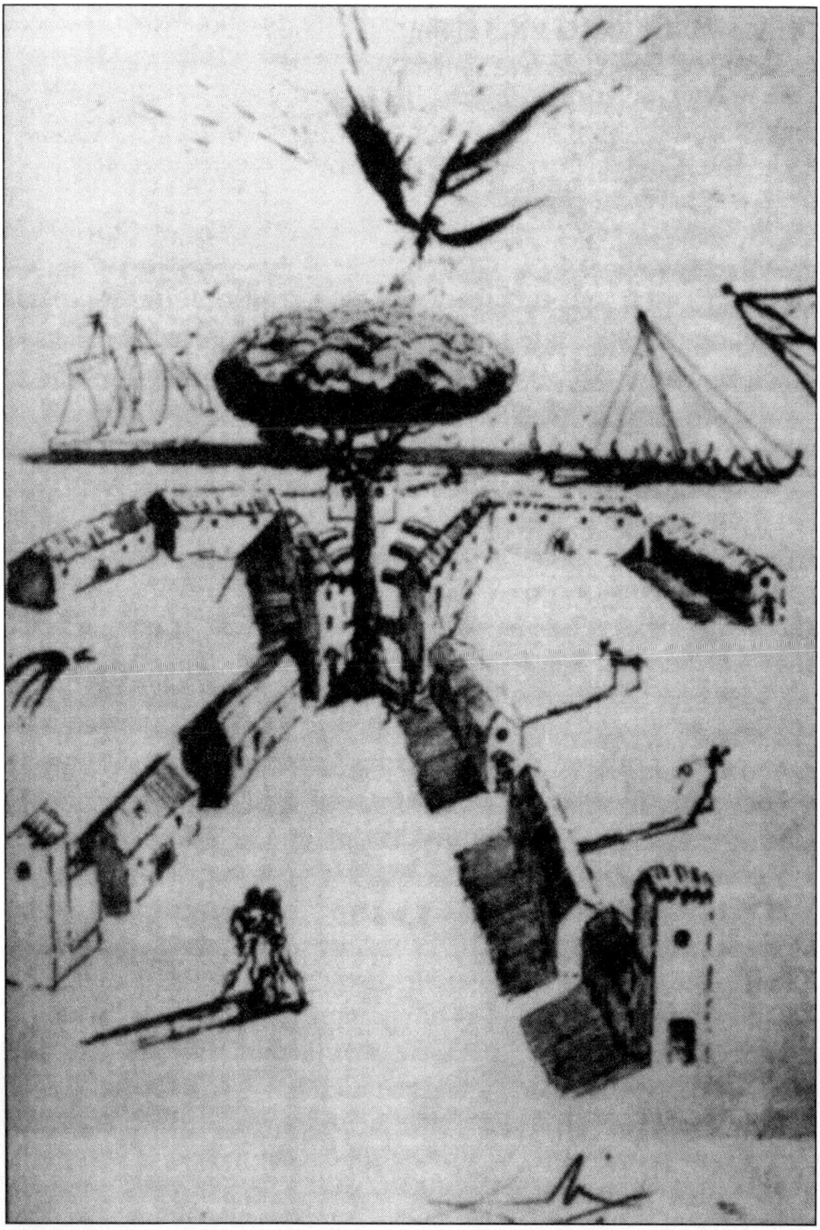

The Beach of Sète by Salvador Dali.

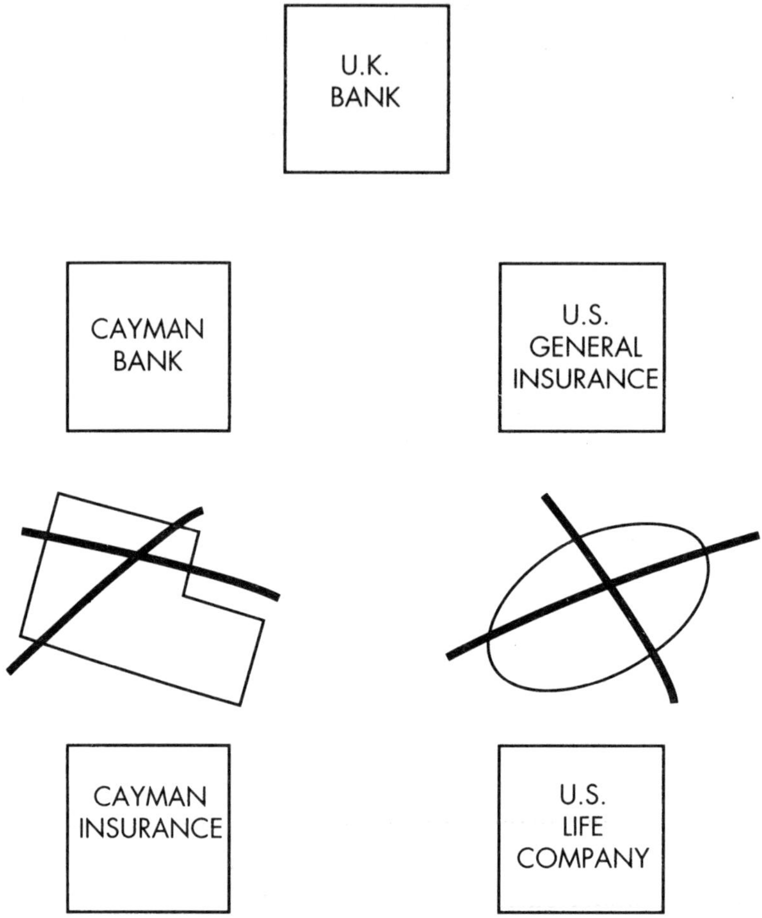

If you get very, very good at all this analysis, you are apt to be called *brilliant,* as in, "That was a brilliant presentation." The original meaning of *brilliant* is something shiny, sparkling, like a diamond. Overlaying this are connotations of rigor and precision. The application of *brilliant* to human mentation was rare before the eighteenth century. Is this an accident, or does it coincide with the Enlightenment? Jefferson talked earlier about the strong "penetration" of Locke, Bacon, and Newton; I think he meant that those men were powerfully analytic. In any event, all that glitters is not gold—or diamond. But one thing is sure, diamonds are tough—we use them to drill through rock—and cold. Sometimes we call them *rocks* and sometimes *ice*. Let's look at temperament.

Temperament

Serious, distant, stiff, intense, no-nonsense, cerebral. The Technocrat is cold. The room temperature drops three degrees when he or she enters. Everyone gets quiet and usually stops smiling. Serious looks get plastered on faces, like when kids are joking around and the third-grade teacher shows up. "All right. No nonsense, children. We're here to work, not to play." They're distant and stiff, and that's why we heard a Craftsman say earlier, "I can't get a grip on them. I don't know what makes them tick." And it's why, after many years, we know them no better than before.

There are thick walls around their emotions. In addition to our conscious thoughts, it's through our emotional contact that we get to know one another. Have you ever wanted to cozy up to *brilliant?* How about taking in a second-rate movie, just for relaxation? Do you invite *brilliant* along? Technocrats are so intense, seem to be under such rigid control, take things, and especially themselves, so seriously that grown adults begin to feel like foolish children around them. So that's why people see them as *cerebral,* which, according to *Oxford,* means "of the brain; intellectual [as opposed to emotional]." To some it may seem inappropriate that I have placed *cerebral* in the temperament category instead of in thought processes. But we all *think.* What is different with *cerebral* is that it excludes affect or emotion. It is just this exclusion that defines the word. In everyday usage, a cerebral person is someone who, paradoxically, telegraphs the emotion that emotion is unacceptable. Why?

Inner Life

In a word, *defense*. Affect—emotion—is at the core of character. If we are frightened, for example, we develop defenses against fear: flight, fight, hide, deny, forget, whistle (you know, whistling in the dark). Depending on our endowments, our choice of defense and the situations we encounter, our character will be shaped: We will develop *characteristic* ways of behaving and thinking; *characteristic* ways of reacting to people, to events, to ideas, to the world. This is the essential reason why many psychiatric texts have in their titles *bio-psycho-socio,* in that order. Now, I'm not going to subject you to a treatise on why these people became the way they are. I know literally nothing about their genetic makeup, their relations with their mothers and fathers or anything else of a personal nature. But I am going to talk about the end result—the *psycho* and its relations with the *socio.*

Here is a typical MMPI profile of a Technocrat as interpreted by the clinical psychologist (remember, he did not know who this person was nor what I thought about him):

> This guy's very defensive, psychically well defended. Not admitting to anything serious. More sensitive than he would like to be perceived. Chronic worrier. Lots churning around inside that's never talked about. There's a flair of creativity but—this is the key—it's repressed and denied. Under stress, he's paranoid (can't trust anyone but himself) and will project blame onto others. He's isolated, angry, tense, cold, rigid, brilliant—like Eichmann.

There's that word *brilliant* again. Sure enough, in the group of 15 men, this man had the second-highest rating on *brilliant.* He also had far and away the highest rating on *distant* and *stiff;* the second-highest on *uncompromising* and *difficult;* zero on *people-oriented, amiable, exciting, warm, generous, easygoing, realistic,* and *open-minded.* Appearances—the adjectives—reflect the inner world. Two technocratic peers said he was *inspir-*

ing; what you see depends on where you sit. They identified with him.[5]

Let's look a little more closely at the profile to try to understand what's going on. "Very defensive, psychically well defended": When we use the word *defensive,* what does it mean? Take the following incident:

> "Did you spill the milk?"
>
> "Yes, but there was so much clutter on the countertop I couldn't help it, and furthermore I've asked you a thousand times not to put the toaster next to the fridge. So it's not my fault."
>
> "Don't get *defensive.* I only meant did you spill the milk or did Billy. I'm trying to teach him to clean up his own messes."
>
> "Oh."

This man is "psychically well defended" against criticism. He assumes that he will be blamed (a bit paranoid), and he develops a host of arguments to defend against this anticipated blame; one of them is to find a way to blame someone else. Sometimes, as in this incident, the blame he anticipates is not forthcoming; his judgment is bad. He *imagines* it. He *projects* the fault onto others because he cannot bear the thought that he is less than perfect. Less than perfect is dangerous, unacceptable. He's very hard on himself and thus very hard on others. In the organizational context, one Craftsman said of him, "His attitude was always 'I found the mistake and I fired the son of a bitch.' It's never his fault." There's an element of narcissism in this too—not narcissism in the popular sense of egotism, but in the clinical sense of *enraged by criticism.*[6] Criticism, explicit or implicit, real or imagined, would tarnish the perfect image he needs to maintain for others, but especially for himself.

"Lots churning around inside." Lots of emotions—anger, joy, sadness—but they're "repressed and denied." The psychologist is using technical language, but what do *repressed* and *denied* mean to us? *Repressed*—consciously eliminated from thought. *Denied*—

unconsciously eliminated from thought, the simplest example being something like:

> "You were supposed to go to the dentist."
> "I forgot."

These people "forget" their lost emotions. The elevation of certain scales in the profile "seems to reflect the aggressive or hostile feelings and impulses that are present to a significant degree, while the scale-three height in turn shows that repressive and suppressive controls are even stronger than the impulse. . . . Although inhibited and moderate, [they] episodically express their aggressive feelings directly and intensely."[7] A Craftsman would tell me of one Technocrat, "If you disagree with him, he explodes on contact." But most of the time they are the picture of serenity. They seem imperturbable. Everything is under control. Just follow the plan and the orders, and everything will be all right.

Following orders is important. One of the highlights of the personality disorder called the *compulsive personality* is that interpersonal relations are organized around dominance and submission. As a subordinate they may seem agreeable, submissive, even obsequious (so they're hard for bosses to detect). Once in power, however, they will brook no opposition and will evince an unreasonable insistence that others submit to exactly their way of doing things; they will show an unreasonable reluctance to allow others to do things because of the conviction they will not do them correctly.[8] But this is intimately related both to the paranoid tinge and to the fear of failure. The script reads something like, "If I'm in absolute control and I'm perfect, then nothing can go wrong. I must make these others submit to my way of doing things. It's for their own good." For the most part, I think, Technocrats usually believe that what they do is for the benefit of everyone. They think their superior intelligence grants them good judgment. To them, the others are wayward children. This is exactly how Frederick Winslow Taylor, the father of scientific management and a notorious compulsive, talked about "the workers." He wanted to "help" them by breaking their work down to the minutest of details, so that conflict and doubt and confusion and disorder would all be eliminated. Workers would make a lot of money, employers would make lots of profit, and everyone would be happy—an industrial utopia. The facts spoke for themselves. His system was foolproof—with the minor exception that workers hated it and didn't want to submit even for their own good. And the odd thing is, this guy Taylor, who wanted desperately to have peace and calm, created nothing but conflict around him and regularly exploded on contact.[9]

Breaking work down to the minutest of details also typifies the compulsive personality: preoccupation with details, rules, lists, order, systems, organization, or schedules to the extent that the major point of the activity is lost. One Craftsman would say of the Technocrats:

> I don't know. Their strategy was very murky. They seemed to be fascinated by tinkering with the thing. You know, systems, structures.

They're big on facts. They're big on organizational charts. Very big on systems and structures and rules. This orientation toward details and systems often means that these people develop technical virtuosity. They succeed based upon it. Of course, it has a price. What distinguishes a personality disorder from a neurosis is that they themselves do not suffer—they think they're just fine—but the people around them do. Technocrats miss the *tone* of social situations because they are too concentrated on the details and facts. Depth psychologist David Shapiro describes the compulsive character (emphasis is mine):

> These people seem unable to allow their attention simply to wander or passively permit it to be captured. Thus, they rarely get hunches,

> and they are rarely struck or surprised by anything. It is not that they don't look or listen, but that they are *looking and listening too hard* for something else. For example, these people may listen to a recording with the keenest interest in, and attention to, the quality of the equipment, the technical features of the record, and the like, but meanwhile hardly hear, let alone are captured by, the music. In general, the obsessive-compulsive person will have some sharply-defined interest and stick to it; he will go after and get the facts—and will get them straight—but he will often miss those aspects of the situation that give it its flavor or impact. Thus, these people often seem insensitive to the *tone* of social situations. In fact—such is the human capacity to make a virtue of necessity—they often refer with pride to their single-mindedness or imperturbability.[10]

They don't know that their orientation toward the facts is part of their character. They lack self-knowledge or personal insight and think themselves reasonable, realistic, and sensible. This lack of personal insight is the other thing that comes out very strongly in the MMPI results for the Technocrats. The MMPI is a tricky test. It contains all sorts of traps. It has *validity indicators* based on some questions which no one should answer *false.* For example, my Technocrats answered *false* to the statement "I get angry sometimes," or to the statement "At times I feel like swearing." Several of the Technocrat profiles were close to invalidation for these kinds of responses. In general, the MMPI tells us:

> Valid elevations in the high to markedly elevated ranges are most likely to be generated by subjects who are honestly describing themselves as they see themselves. They tend, therefore, to be overly conventional, socially-conforming and prosaic. Some of the descriptions actually correspond to their habitual patterns of behavior while other features of their test answers reflect their poor insight and limited self-knowledge.

They really do not see themselves. Thus, while they think themselves realistic, no one else (except maybe another Technocrat) does. And they see themselves as honest, but no one else does. In an interview with a Craftsman, I noticed that he had failed to mark *honest* for any of the Technocrats. I asked him why. He replied, reluctantly, "I suppose they're honest, but they're certainly not straightforward." It's hard to be honest with others if you don't know yourself, and this absence of authenticity makes itself felt. We will see this more clearly later.

One of the best ways to deny your feelings, and it's the preferred strategy of Technocrats, is to cut your feelings off from the thoughts that go with them. Professionals call this *isolation.* It shows up in typical fashion in this Technocrat's description:

> Oh, him, he's a great guy. I like him. We still have lunch together. He just couldn't do the job, that's all. He understood this. It was nothing personal. Just business.

You see, there is no emotion, no affect here. Just thought. Night follows day. Facts are facts. No hard feelings. Needless to say, the "great guy" he was talking about despises him. A very close cousin of isolation is *intellectualization,* which "represents an attempt to avoid objectionable impulses or affects by escaping from the world of emotions into a world of intellectual concepts and words."[11] That's why Technocrats sometimes sound like talking machines. They have lots of high-blown theory that bears little relation to their actions. One Craftsman had this to say about one of them:

> He used to quote Drucker to us all the time, saying that you could have centralization or decentralization but nothing in between. So he said he was in favor of decentralization. But, of course, he was doing just the opposite.

He always justified, rationalized, intellectualized his actions, usually by relying on some "expert" (read: A guy from out of town).

An Artist said of another Technocrat:

> Oh, sure. He was always talking about decentralization, but every single move he made was centralizing. Everybody knew it. It [the rhetoric] was just for the parade.

Everyone thought the Technocrats were habitual liars (which is why no one thought they were honest), but I'm not so sure they were consciously duplicitous. My impression is that Technocrats sincerely believe their conventional ideas—if Drucker says it, it must be right. They simply can't help themselves. They are strangers to themselves as much as they are to others. Their need to be in control, to dominate, to prevent mistakes, automatically causes them to act in a way that removes authority from everyone else. As we will see later, they intellectualize this process with the fanciest of reasoning.

So. With their emotions securely buttoned-up, except for rare explosions, and with grand intellectual schemes to impose, they come across as cold, rigid, and brilliant. Others mistrust them because underneath their protestations of kindness, underneath their mouthing of the importance of teamwork, lurks the uncompromising hostility and self-righteousness that everyone dreads—"It's my way or the highway." Fearing the unknown and the uncontrollable, logic, research, and rational analysis become their passwords, because these things promise to control and to tame an otherwise frightening and uncertain future. Under stress, they project their forgotten emotions onto others, calling them emotional and impulsive. They blame others for the errors an uncertain world inevitably gives rise to. Once they have ordered their world, Technocrats are stubborn about changing their views. They are conservative because change is deeply menacing to them. This is part of their character.

Can you teach someone who is serious, detail-oriented, rigid, and methodical to be imaginative, wise, entrepreneurial, or funny? If so, you're a better teacher than I.

The Technocrat Character

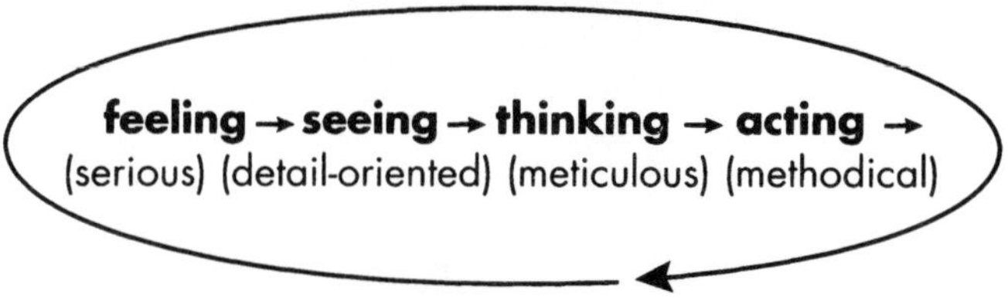

So far, I have given you the three characters in relatively splendid isolation. In the next chapter, we're going to see them all together. In the chapter after that, you'll see them interacting on the organizational stage.

4

Behind the Scenes

Scientific statements are, as people say now, far more easily "cashed." But the poet might of course reply that it always will be easier to cash a cheque for 30 shillings than one for 1000 pounds, that the Scientific statements are cheques, in one sense, for very small amounts, giving us, out of the teeming complexity of every concrete reality, only the "common measurable features." Such information as Poetic language has to give can be received only if you are ready to meet it half way. It is no good holding a dialectical pistol to the poet's head and demanding how the deuce a river could have hair, or thought be green, or a woman a red rose. . . . If he had anything to tell you, you will never get it by behaving in that way. You must begin by trusting him. Only by doing so will you find out whether he is trustworthy or not. Credo ut intelligam. . . .

C. S. Lewis, *Christian Reflections*

Given what you know now, do you think the Technocrat could trust the Artist long enough to find out whether or not she is trustworthy? Not on your life. Do you think he would try "holding a dialectical pistol" to the Artist's head? Try to force her to define the undefinable? You're right.

Well, that's just a beginning. What we're going to do now is put those little circles I gave you in the preceding chapters into perspective. The circles, the characters, exist in relation to one another and not in the artificial, analytical isolation in which I first presented them.

Following you'll see the whole map. I have superimposed the circles on it so you can follow the cast of characters. We talked first about the Artists. Here they are again, next to their friends, the Craftsmen, and literally opposite their enemies, the Technocrats. Look at Cobb. You might think that he is most different from Brien, Brien being at the opposite end of the spectrum. But here's the trick: He's almost on the same level as Brien on the horizontal plane; he shares something with Brien. The person Cobb is most unlike is Cam. Cam is in the northwest, Cobb in the southeast. Both dimensions count. Got it? This is how we're going to proceed.

There are multiple levels of interpretation of this graph, so I hope you'll be patient. First, it can be read like an organizational thermometer. Start in the extreme east and it's very hot, about

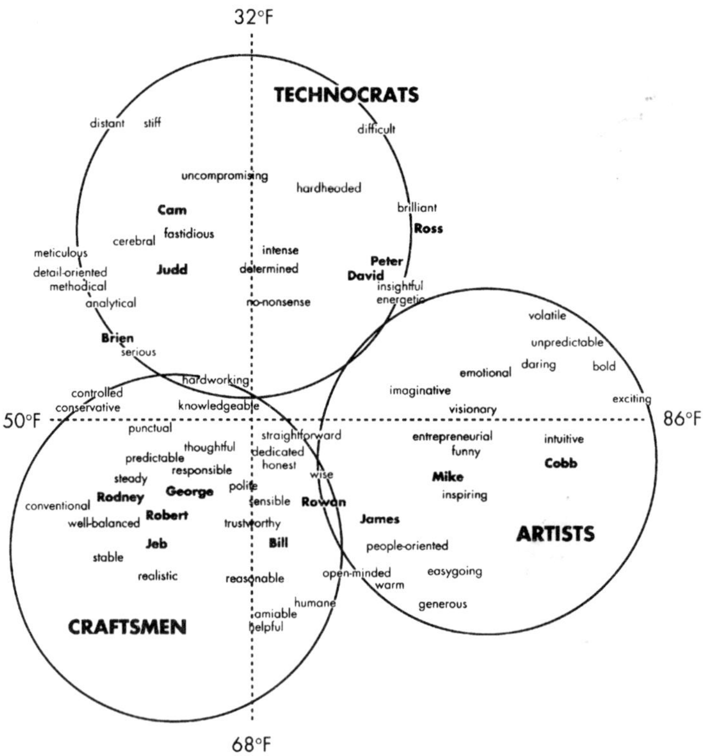

86°F—exciting. The organizational melody is like a concerto with one key guy—Cobb, Mike, or James, take your pick—leading it. When you're inside the organization, everyone is working diligently and everyone seems happy. Some are breathing hard because things are really hopping. A dozen crucial projects are under way. They're attacking on all fronts.

As you move southward, the temperature gets milder, more temperate, around 77°F. You pass from hot to warm. From Cobb to James, you're moving from the Artist territory into the Craftsman territory. In the fancy arithmetic I mentioned earlier (also see Appendix C), James loads mainly on the Artist, but partially on the Craftsman. Moving due west, you pass through Bill and Rowan, who have some artistic leanings but are basically Craftsmen. In the heart of the Craftsman profile we pass from warm into neutral territory. It's now a very comfortable 68°F, and everyone goes about their business at a straightforward, relatively calm pace. Relations are cordial but not effusive. There are lots of smiles but little uproarious laughter.

As we round the cape and start to move northward again, past Rodney and reaching up to Brien, there's a distinct chill in the air. Let's say the workplace is dull; nobody's particularly enthusiastic but neither are they comatose. We're in serious territory. Pushing past Brien up to Judd and Cam, this is Chicago in January. Stiff, distant, uncompromising, cold. Brrr. Inside the organization, it's like a morgue. No one talks about anything except business. Chatting is frivolous. The corridors are silent. All the office doors are closed and you just know there is very, very serious work indeed under way behind those closed doors. There are an equal number of projects here as there are in the Artistic organization, but they all have to do with financing, with systems, with internal and external audits.

As you round the top and start to move toward the territory of David and Peter and especially Ross, it starts to get hot again, but it's the heat of a blast furnace and not of a tropical sun. That's not a very kind image, you say. Well, it's not meant to be. I'll tell you why, because this moves us to the next level of interpretation.

It's this business about the horizontal and the vertical axes and their relationships. The heat of the Artists is tempered by their affinities with the Craftsmen. Cobb, Mike, and James are all *below* the horizontal axis. Ross, on the other hand, is well *above* it. All the people up north are radically non-Craftsman. That is, they are not seen to be, among other things, honest, trustworthy, amiable, easygoing, or generous. In fact, they are only hot in the sense of intense, demanding. They are impatient and they often explode on contact. The more positioned to the right of the vertical axis, the more emotional, but the emotions up north are not pretty ones. In fact, they can be downright ugly. Thus, when people tell me that the secret to great leadership is authenticity and letting out what's inside, being spontaneous and all that, I get nervous. I don't particularly want these guys to be spontaneous. I would just as soon they kept their emotions to themselves—or confided them to a therapist. Their repression is good—for the rest of us.

So let's get back to Brien. Brien does not usually explode on contact. But he, and all those positioned in the west—Rodney, Brien, Judd, and Cam—are radically non-Artist. That is, they are radically *not* intuitive, inspiring, entrepreneurial, or funny. They are instead controlled, conservative, methodical, and cerebral. If we look at Rodney, for example, he's a long way from Cobb but an almost equal distance from Cam, even though he's methodical like Cam. You would think he wouldn't get along with either of them. Wrong. He gets along very well with Cobb. Let's try to understand why.

Imagine you are sitting down in the southeast corner close to James. When you look diagonally across the map you see your opposite. You see cold, cerebral. You see Cam and Judd. Do you hate them? Okay, now move across and sit in Cam and Judd's shoes and look back at James—intuitive, funny, inspiring. Are you inspired? Do you think he's funny? Think carefully before you answer. Remember, you are now stiff, methodical, and serious. There's one clue, one key word here that makes all the difference in the world. Their reactions to each other, the Artist and the Technocrat, are *asymmetrical.* The Technocrat despises the Artist. The Artist tolerates, even appreciates, the Technocrat; as we will see later, one of our Artists will give a Technocrat total control over his organization. The key word is *open-minded,* with a minor in *generous.*

When Artists look at the Technocrat they don't hate him or her. They see *brilliant.* They see *methodical.* They see his or her qualities and virtues. They find the Technocrat sort of stiff, but they're tolerant of it. They want the *brilliant,* for they want to be challenged. They don't like yes-men. They want to be stimulated. Their minds are open. And they're a bit insecure about their own skills. They have doubts. Their way of protecting themselves is to have very strong people around them. People who have the courage to tell them when they're going off the rails. That's one of the many reasons they value the Craftsman, for the Craftsman is direct, frank.

But when Technocrats look at the Artist, they are not open-minded. They see only faults. Here's what one Technocrat says:

> He was *Star Treky.* He produced a hundred new ideas, but only one of them would be useful. It's terribly inefficient and ineffective. The people underneath get *frustrated.* It's awe-inspiring but eventually demotivating. Our world cannot do without them, but our organizations sure can.

Another adds:

> We had to protect the organization from him. *He drove people crazy with his ideas.* No one knew what he should be working on at any one time. Everyone was running around in all directions.

And still another:

> He simply was not a manager. He was over his head. He was an adviser. He was volatile, unpredictable, imaginative, visionary. A nice guy. Very amiable. The kind of guy you might seek ideas from but not the guy to run something. He *didn't know how to work.*

This is what happens when *serious* looks at *funny*. It sees red. "He didn't know how to work." Translation: He was not serious; he took vacations. "He drove people crazy." Translation: He drove me crazy. "The people underneath get frustrated." Translation: I got frustrated. Others in these organizations were much more apt to tell me things like, "You work for a guy like him only once in your life." That was said with deep admiration. As for the guy who "didn't know how to work," a senior member of his board told me, "He was very well mannered and considerate. He fired people, but he did it nicely (if firing can ever be nice). He was *hardworking* and very *dedicated.*" I didn't hear any frustration except from the Technocrats. The Craftsmen weren't frustrated at all.

On the contrary, Craftsmen respect the Artist. Oh, they think he or she sometimes goes too far and too fast, but they think Artists are indispensable; one told me, "The analytic boys are a dime a dozen, but you can't buy dreams." Yet Craftsmen, normally open-minded, shut down when it comes to Technocrats. They're the ones who get intransigent, stubborn. This is essentially because the Craftsmen have their feet most firmly planted on the ground; they don't mind being lifted off occasionally by the Artist, by a dream, but they resist with all their might getting carried away by a theory. They find the Technocrats, and their managerial theories, their maxims, totally unrealistic. This is what Craftsmen say about Technocrats:

> They seemed to think that you could grow and expand and not have an impact on the bottom line. But any schoolboy knows better.

> They're like the little old lady who wants capital gains and income. Totally *unrealistic.*

> You can't correct a problem unless you see it exists. It's like me. I look in the mirror and I see a young fullback, not a balding, middle-aged man with his chest on his belly. You have to see *reality* to change it.
>
> Why can't they see it, they're brilliant? I guess if, with all their brilliance they can *rationalize* it all away, we all rationalize.

> They didn't really understand our situation . . . *too distant from the coal face* to understand how the business works.

"Too distant from the coal face," that is, too theoretical, impractical, not experienced enough to know the difference between a good idea and a bad one, a new idea and an old one dressed up in a pretty new skirt. Craftsmen value experience, hands-on experience. They don't believe that you can get to know something only by reading about it. You have to be there. That's why they think the Technocrat makes a good consultant but a lousy manager, and even if he or she's their boss, they say so.

Technocrats, on the other hand, don't value experience. For them, experience equals old-fashioned. They value brain power. Here's a taste of technocratic opinions about Craftsmen:

> He can be a useful number-two man. He can be counted on to keep the ship afloat in stable waters. Once the direction is set [read: by us], he can keep it running for a while. But after a time, the disease sets in. Inefficiencies. And when the waters get troubled, when the marketplace is fluid, he's hopeless . . . he doesn't have the *brain power.*

> We have to put some *talent* in there. He's just not strong enough.

> I have a feeling that he's over his *head.* Not really up to general management. Really a second-level kind of guy.

> It's not that he's stubborn. He's just too *stupid* to follow the line of reasoning.
>
> They're *incompetent.*

Hear the disdain, the condescension? Some of these "incompetent," "stupid" Craftsmen had been running businesses while many of the Technocrats were still in high school. Now, experience is not always profitable. Some people *are* old-fashioned and closed-minded. But every one of them? Don't forget that these Craftsmen are seen to be open-minded by the vast majority of their colleagues. They resist change all right, but only when the change is for them unrealistic.

The irony is, of course, that the Technocrats see themselves as realistic, sensible, and reasonable, but they are all alone. They also see themselves generally as entrepreneurial, intuitive, and imaginative, but they are all alone. One described another to me as an entrepreneur. Because he was the only one to say so, I was curious about his definition of entrepreneur, so I asked him to elaborate. "An entrepreneur," he said, "is someone who is stingy, someone who treats the corporation's money as if it was his own." That was the most bizarre definition of "entrepreneur" I had ever heard. No wonder, then, that another Technocrat told me:

> James, Cobb, and Rowan, they were all dreamers not entrepreneurs. They're what I call advisers. Now we have builders. They're conservative of course, but if you're not conservative you'll lose everything.

This takes us back to the problem of personal insight. Technocrats lack insight into their own characters and into others'. I very nearly burst out laughing when the coldest, most rigid of them said of another, "Him. Oh, he's a very cold man. Frigid. Very distant." The person he was talking about was the life of the party next to him.

Even within the group of Technocrats, there are problems. One of them referred to his colleagues as "a bunch of amateurs."

Others felt that some of their peers, their rivals for power, were not really "professional" [read: technical] enough. Of course, if you possess *Truth* and you need to impose *your* truth, everyone is a dangerous rival. You have to find some way to undermine the other's credibility. Those who are outwardly calm and collected are suspicious of anyone who is more emotional; they say things like, "He'll have to be watched and controlled," or, "He can fly off in all directions." They are really deeply suspicious of *emotional.* In fact, they use it as the worst kind of insult. Of course, anyone who needs desperately to be in control will have a hard time working in the traces of a team, no matter what its composition.

So, inside the organization, behind the scenes, we hear a cacophony of people shouting insults at one another: "emotional," "unrealistic," "dreamer," "theoretician," "incompetent," "old-fashioned." They think they are arguing about substance, when in fact they are arguing about character. Each sees the organizational landscape differently. Each interprets it differently. The same facts have different meanings. One is buoyantly optimistic and occasionally discouraged. One is moderately optimistic. One is deeply pessimistic, even phobic. Whose interpretation will carry the day?

The Technocrat's. Let's watch how he steals the scene.

Part 2

The Play

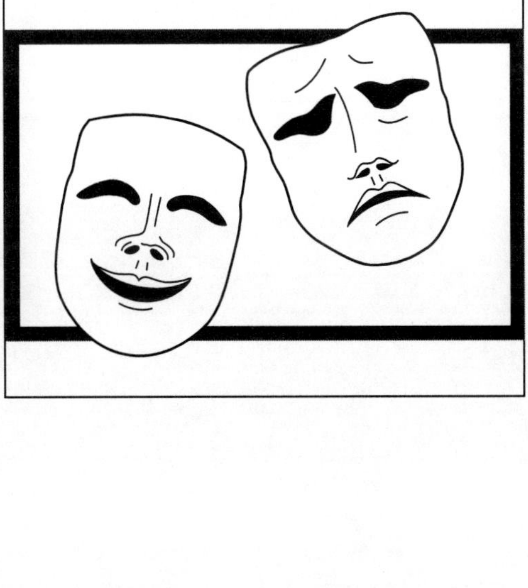

5

The Dreams of the Playwright

There is, however, no basic vocabulary of lines and colors, or elemental tonal structures, or poetic phrases, with conventional emotive meanings from which complex expressive forms, i.e., works of art, can be composed by rules of manipulation. . . . It is easy enough to produce standard cadences, manufacture hymn tunes according to familiar models and some experiential knowledge of standard alternative resolutions, etc.; but such products are at best mediocre. . . . The analysis of spirited, noble or moving work is always retrospective. . . .

S. K. Langer, *Mind: An Essay on Human Feeling*[1]

We have seen that there have been three Ideal Types of leaders involved in the life of the organization. The Artist appears volatile and imaginative, corresponding to an inner life that is cyclothymic (up and down) and autistic (the capacity, or indeed the need, to dream).

The Technocrat, on the other hand, is meticulous, cerebral, and, often, brilliant, as well as difficult, uncompromising, and emotionally cold. These are the outward manifestations of inner psychic processes: fear of the unconscious and the emotions, which causes a generalized suppression of all emotion along with a disdain for it in others. These fears also engender an unreasonable attachment to the security afforded by conventional wisdom: If everybody thinks it's true, it must be. This need to intellectualize, coupled with above-average intelligence, makes the Technocrat a powerful analyst and a dangerous foe.

The Craftsman is not simply midway between the two, a residual category like "all others." The Craftsman is honest and dedicated, loyal and straightforward. The Craftsman is intelligent, forceful in defense of personal values, with an appreciation of hands-on experience that sets him or her apart from fellow colleagues. Psychically, the Craftsman is more than Everyman; he or she has above-average drive, otherwise he or she would not have sought positions of power. The Craftsman is polite but not a pushover, and refuses to hide personal values in order to survive.

I'm sure you can already imagine how these three characters would interact on the organizational stage. This chapter retells their story from the far from dispassionate point of view of the actors themselves. The names and places are fictitious, but the events are not.

The Early Years

Our story opens in Small City, United Kingdom, in the 1960s. The aging chairman of the board of a 40-year-old, medium-size, general insurance company—ABC Company—is preparing to retire. He does not have confidence in the ability of any of his employees to take over the managerial reins. Recently he has come into contact with James (our Artist and chief protagonist), the young owner of a small rival company, and was very favorably impressed with the man's style. Self-taught, James seems to have a sophistication beyond his years and to have established a reputation as both scrupulously honest and farsighted. The chairman courts him. The younger man is not anxious to give up his entrepreneurial freedom. Finally, after a great deal of soul searching and with the prospect of much wider horizons before him, he accepts. Within a year, James is firmly in control. Within 18 months, he makes his first acquisition. For the next 10 years, the aging chairman cannot catch his breath. He never regrets his decision.

In 1965 ABC Company had a small but important hold on its regional market. A local favorite son, it benefited from its reputation for strengthening the local economy (indeed, that was its raison d'être); "home rule" sentiment was running high. Whereas previous management had been content to tap this local sentiment, James felt that more could be done; the whole of the country could benefit from the success of a local firm. However, at normal growth rates, ABC Company would never overtake its much larger foreign-owned competitors. "If we and company X had both grown at the same rate per year," James figured, "then they would have added two hundred million sterling to assets while we added only ten. They would have got farther and farther ahead of us. That's why we had to acquire."

With a limited capital base, a property and casualty company is restricted in its capacity to make large acquisitions. James, now the chairman, decided to solve his dilemma in two ways. During the course of the first acquisition, a troubled local insurer, he had come into contact with the large European Financial Group (EFC), which had held a controlling interest in that firm. James had negotiated a sweetheart deal with EFC that brought that company both peace of mind and steady dividends on its remaining minority interest. James' personal charm and astute negotiating skill in not taking advantage of EFC's momentary weakness had earned him permanent—and wealthy—friends. He created a strategic alliance *20 years before* academics started talking about such things. He was to use this alliance repeatedly over the next decade, and his formula for controlling a target with less than 50 percent of the stock was to allow him to leverage ABC Company's limited assets far beyond their paper worth.

The second axis of his strategy was to buy a life insurance company. Because life insurance contracts and liabilities are very long-term compared with general insurance, a life company builds up enormous assets. Controlling those assets gave James a much larger pool of capital to play with, and play he did. He bought a real estate company. Taking over a holding company in order to get his hands on its insurance assets, he acquired shares in a small bank and a building society. Believing that the future belongs to those who see it coming, he bought modern communications companies, companies that might just end up providing the best way to sell insurance. (And this was *25 years before* anyone started talking about the Electronic Highway.) While his competitors sat smugly and safely behind the walls of their sectors—banking, general insurance, brokerage, life insurance—he broke out of the industry recipe and out of the borders of the financial sector as a whole *20 years before anyone else.* Personally indefatigable, with a restless, effervescent energy, he was never content to rest on his laurels. In his home country, he was third in general insurance, sixth in life; there was much to do.

Known by then as a man on the rise in corporate circles and because he had a reputation for moving decisively when opportunity presented itself, many prospective deals were brought to his attention. In 1976, he was on an airplane with an acquaintance whom he knew to be dissatisfied with the performance of his British life company. He whispered in his ear that if ever this gentleman wished to sell, he would be interested. The call came shortly thereafter. With some fancy legal footwork, and the assistance and participation of the ever-present EFC, he managed to swallow the London-based life company, which was many times his own company's size. Jubilation was the order of the day. Not only did he now have an important presence in a major world financial center, but the life company had subsidiaries in Hong Kong, Canada, and the United States. In one sure move, he catapulted the previously obscure ABC Company onto the world stage. To the outside world, it looked opportunistic; to him, it was all part of the design.

The following year saw him take over another London holding company, which allowed him to solidify his position in general insurance and pick up more shares in the bank and the building society. In the process, he got more real estate for his real estate development company and a woolens mill. A decentralist by nature, he gave free rein to the local management in Canada and the United States, and assets began to climb there as well. Finally, by the end of the first five-year period, he controlled the bank and was able to appoint its director. It was now 1980. He'd been in power for 15 years. For him the time seemed short.

His empire was big ($2 billion in worldwide assets). He couldn't run it all. At the time of the British life company acquisition, he'd felt it necessary to hire a very senior person to oversee it, someone with more life insurance knowledge than he had. And he had to put someone in the bank. He continued to watch over general insurance, considering himself very ably seconded by an operating executive he had picked up and promoted during an earlier acquisition. The other disparate pieces were also his responsibility. Thus, by the end of the decade, the organizational structure and the people in it looked like the following chart. James was at the top and is, as we've seen, an Artist. Running the life company was Cameron, a Technocrat; its U.S. and Canadian subsidiaries were run by an Artist, Mike, and a Craftsman, Rowan. There was a Craftsman, Robert, at the bank; another one, Jeb, was in general insurance; and a third, Rodney, had miscellaneous responsibilities.

The Man, the Organizational Climate, the Strategy

What kind of person was this who was able to build an empire with few resources and in defiance of the conventional wisdom of the day, when bankers and insurance executives had the reputation of

Structure as of 1980

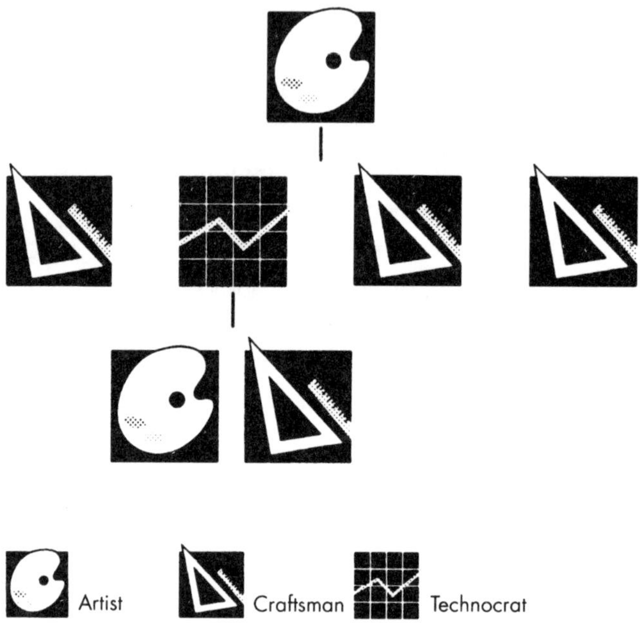

being, at best, stuffy? Tall, strong, and handsome, with the bearing of the Scottish king after whom he was named, James easily conquered all those he met. A board member told me:

> When he walked into that London boardroom for the first time after the takeover, everyone was on guard. Suspicious of the "upstart," these men expected the worst. They expected him to be both mediocre and arrogant.

By the end of the day, he had allayed both fears. He conceded that the Londoners knew more about their business than he did. He reassured them that he wouldn't interfere. He shook hands all around. He wined and dined and toasted their collective futures. He left having won them all over. What kind of magic did he have?

He had the magic of the Artist. Remember the description of James in Chapter 1? The sentiments about him were almost universal: *imaginative, intuitive, visionary.* A 100-percent score on *energetic* and *entrepreneurial.* These qualities made him an inspiring leader, a leader who gave people a sense of going somewhere and of participating in something important. He had a real gift. He was always talking in images and metaphors about where ABC Company would be in 20 years, although he never put it on paper. As well, he had a prodigious memory. During heated negotiations, he didn't need to refer to financial crib sheets. He remembered all the numbers. But talent alone can sometimes wear thin, and as important as these qualities were, it was perhaps his other side that kept people close, loyal, and dedicated.

His other side was *craft.* Everyone, even his enemies, described him as people-oriented, wise, honest; 9 out of 10, as humane, amiable, and warm. If you remember our map of the players, you see James positioned halfway between the Artists and the Craftsmen. And although he is a polar opposite of the Technocrat—appearing in the southeast opposite the Technocrats' northwest quadrant—his craft virtue of open-mindedness made him tolerant of other personal styles. He could work well with Artists, Craftsmen, and Technocrats. He admired the Technocrat's skills. Self-taught himself, he had a tendency to be in awe of the formally trained. He scored zero on *distant, stiff, cerebral, difficult,* and *uncompromising.* Unfortunately, for him and his organization, he also scored zero on *meticulous* and *detail-oriented,* and far too high on *emotional.*

The pure Technocrat, lacking the Artist/Craftsman's open-mindedness, found him beneath respect. James' vague images of a distant future struck the Technocrat as ridiculous, even foolish. James' passion, his dedication to social, economic, and organizational causes, made him appear sentimental. His decisiveness and impulsiveness were anathema to those who felt that management was a very serious affair, something that necessitated knowing exactly where one stood, having all the facts documented and the strategy clear. One told me with scorn in his voice: "Woolens mills, TV escapades. He didn't know what he was doing. It all had to be cleaned up. Put into order."

The metaphors and images that James' admirers found inspiring, a Technocrat detractor called *riddles* that sometimes made no sense even after "decoding." This is always the Artist's cross to bear; his visions are often inarticulate. Remember James's version of the Dali painting in Chapter 1? He was accused of not knowing what he was doing because the design was in his head, not a plan on paper. James says, in his own defense, "They accused me of diversifying, but I wasn't diversifying. These things were all related; they just didn't know it." But, I asked, what about the woolens mills and the other bits and pieces? "I wasn't interested in any of that stuff; I wanted the underlying assets. The other stuff was all sold at a profit."

If this lack of clarity was not enough to tarnish his reputation, his emotions were. One Technocrat said, "He's incapable of making a *logical* decision; he's too emotional; he cares about where the head office is located; he doesn't understand that it's just a building; it's necessary to move. It makes economic sense." (The Technocrats wanted to move the head office from Small City to London; James insisted that he had a moral commitment to stay. The Technocrats would eventually win.) Another added, "He was incapable of firing anyone. He was too close to his people. He always lets his *feelings* interfere with his judgment."

This was his Achilles' heel. It would prove crippling.

For the moment, however, he was still firmly in charge. His messages to shareholders in annual reports spoke to his central preoccupations: growth, the creation and maintenance of a corporate culture that enlists the creative outpouring of its employees, and social progress for Small City and for his country. Poems introduced the messages. In one, he almost apologized for having to report on earnings:

> Even though we don't believe an institution like ours should be judged strictly on the basis of the judicious and efficient use of resources, nevertheless, once a year we have to tally up. . . . Enough for numbers. . . .

He very quickly went on to quote Saint-Exupéry on the importance of being larger than ourselves, carrying within us the hopes and dreams and well-being of others (20 years before anyone talked about empowerment). Speaking of the corporation's reason for being, he talked about friendship and love:

> Man reacts instinctively to the pressures of material need and the desire to live in friendship and love. It seems to us that to respond to these twin needs gives our institution a worthy reason for being.

To the Artist, the organization must root itself in moral imperatives to justify its existence.

James was not afraid of the future; with respect to its dangers, he wrote:

> Our era demands much but promises even more! Even though we may feel a certain vertigo in front of the dangers, we can also taste—even exult in—the joys of knowing, creating, living.

Often philosophical, he dared to write publicly that aggression, animosity, even hatred are nothing but "an unconscious cry, an urgent appeal for more love." As though to underscore his point, he blanketed the cover of the annual report with the names of hundreds of his employees. Harvard didn't tell him to do this. He did it instinctively. Finally, amidst all this talk of harmony and understanding, was always the note on growth. "An enterprise," he wrote, "must always be judged by its investment in the future and by the *ingenuity* with which it acts as a *pioneer.*" As Polanyi tells us, "The explorer's fumbling progress is a much finer achievement than the well-briefed traveler's journey." And this was not growth for growth's sake, but growth in order to accomplish great things for his country, for its citizens. Sentimental? Some thought so. In any event, as the next period opened, both he and the company he led were in full flight.

6

The Harsh Realities of the Stage

Blow, blow, thou winter wind,
Thou art not so unkind
As man's ingratitude:
Thy tooth is not so keen,
Because thou art not seen,
Although thy breath be rude.

Shakespeare, *As You Like It*

The first two years of the 1980s were marked by the consolidation of the bank and another major acquisition in general insurance. Already, corporate assets were more than $2 billion. James had been very busy. But another preoccupation began to weigh on his mind. Although still very vigorous (he still is at 75!), he was in his mid-sixties and felt he must plan for succession. Long walks and long talks with his wife were the order of the day. Never arrogant, he told himself, "Sometimes one can stay too long in an organization; it needs fresh air, a new approach." Anxious to provide ABC Company a secure future, he had doubts about his own continuing vision; he'd been taught, falsely, that vision is a characteristic only of the young. In his mind, there was only one person for the job, only one person with the breadth and the knowledge to understand all the pieces of the puzzle that had become his empire. This man had been with him since a major acquisition in the 1970s. He had the confidence of the board and he was all that James felt that he himself was not: brilliant, trained, solid, serious. In 1981 he invited Cam to accept the post of managing director. James would stay on as chairman. The offer was accepted.

To free both of them for other responsibilities necessitated hiring senior people for the two largest operating divisions. James and Cam together interviewed the candidates and jointly agreed upon two of them. The Artist, James, continued in the chair and, effectively, remained very influential, with his power base rooted in the shareholders and EFC. A Technocrat, Cam, became the number-two staffer, and as fate would have it, the recruits turned out to be a Technocrat, Judd, and an Artist, Cobb.

As the number-two staffer, Cam had already become more and more influential. James allowed himself to be persuaded to sell off the "bits and pieces" that, according to Cameron, "made no sense." James's dream of distributing financial services through electronic means had not come to fruition soon enough. According to Cam, it had been a pipe dream (time and the Electronic Highway will tell). The communications interests were sold off. In fact, anything that was not, strictly speaking, financial services was sold off. Two new sectors were added by acquisition: investment banking and investment advisory companies. By now, the whole world had begun to talk of the collapse of the "four pillars": banking, trust, insurance, and brokerage. It had become the new industry recipe to be involved in all four and, where regulation permitted, to engage in cross-selling and marketing. *Synergy* became the new watchword.

Meanwhile, the two new people, Cobb and Judd, were settling into harness, and already things were beginning to heat up in the largest sector. Cobb, the Artist in charge there—seconded by Rowan (a Craftsman with entrepreneurial tendencies) in the United States, and by Mike (another Artist) in Canada—was already on the march. In the spring of 1984, Cobb approved a life company acquisition that would more than double U.S. assets. Similarly, in Canada Mike was giving vent to his acquisitive instincts, buying up life companies and blocks of business. By 1985 the parent company's need for capital was seemingly insatiable. Public issues and private placements took place literally monthly. With the acquisition of the final controlling block of the bank's shares, the company was now able to remove the fiction of equity accounting and consolidate the assets of the bank into its balance sheet. Assets were ballooning, and it was a propitious moment to use capital markets. The company was on a roll; it seemed in good hands; the team seemed to work. James decided to step down and to transfer all power to the number-two person, Cam.

I interviewed James shortly after he promoted Cam. "Is there anything," I asked, "anything that could jeopardize the future of your firm? Could it be taken over by bureaucrats, for example?" He replied, "No. Its entrepreneurial character is too well entrenched. It's too strongly ingrained in all the people."

Nothing changes overnight, but changes were to come.

Changes: 1985 to 1990

So James disappeared from the power structure and became increasingly isolated. He remained on the board, but both formal and informal power were now in the hands of another. What kind of person was Cam? As noted earlier, he was a Technocrat. On the map, he's located in the northwest quadrant, opposite James in the southeast. His lowest scores are on *humane, amiable, warm, helpful,* and *generous.* James chose his opposite to succeed him. Why? Well, for his talents. Cam is serious, cerebral, hardworking, meticulous, determined, and methodical. He scores the second highest of all on *brilliant.* He's conservative, controlled, intense. He seems very sure of himself. Not a man to make many mistakes. Solid, even stolid. Inspires confidence that everything will be under control, everything in its place. No loose ends. Rational choices. Analytic procedures. Sophisticated, in the way that the self-taught may find themselves deficient. When asked why James might have made such a choice, one former executive vice president stated simply, "He made a mistake; human beings do, you know." James would later rue the day.

The Team: 1986

It is appropriate, now, to pause to get to know the rest of the team, because it was during this brief interlude that a team actually existed. Although the formal power structure saw each of the four group CEOs—Judd, Cobb, Rodney, and Robert—reporting directly to the chairman, Cam, the atmosphere of power was much more diffuse. Each of the four held a seat on the nine-member board. James, the previous chairman, was still a member and continued to carry moral weight. The small board functioned with little ceremony, more like a management committee than a formal board of governors. Everyone had a say; there was no obvious deference to hierarchical authority. Deals were brought to the table and consensus achieved with little ado. If there was deference at all, they each deferred to one another's right to proceed independently in their sectors. Everything was going well financially; there was little tension in the group.

Judd, one of the two senior staffers James and Cam had hired, is a Technocrat; on the map of the three circles on page 64, you can see that he is positioned within millimeters of the new chairman, Cam, in the northwest quadrant. On the checklists, no one calls him *volatile, funny, exciting, visionary, intuitive,* or *inspiring.* One calls him *entrepreneurial.* He is quite clearly not an Artist. No one calls him *honest, humane, people-oriented, wise,* or *open-minded.* He is clearly not a Craftsman. On the other hand, a majority calls him *controlled, serious, intense, analytical, methodical,* and *determined.* At board meetings, he talks of the progress being made in the merger of his various subsidiaries and the structures and systems that will result from it. He shares with us glowing profit forecasts based on the *trends.*

As we see in the following organizational chart, reporting to him is Jeb, an archetypal Craftsman; on the map he's in the middle of the craft pack. At the top of the checklists on him are *honest* and *well-balanced,* followed immediately by *humane, amiable, sensible, dedicated,* and *knowledgeable.* He's also *predictable* and *conventional;* both admirers and detractors agree on these adjectives. Never cited as *difficult, stiff,* or *hardheaded,* he has relationships that are generally harmonious. He is respected and liked by his subordinates, appreciated by the Artists and other Craftsmen, and discounted by the Technocrats; at the time, they say things like, "He's weak. We need to get rid of him," or, more

Structure early 1980s

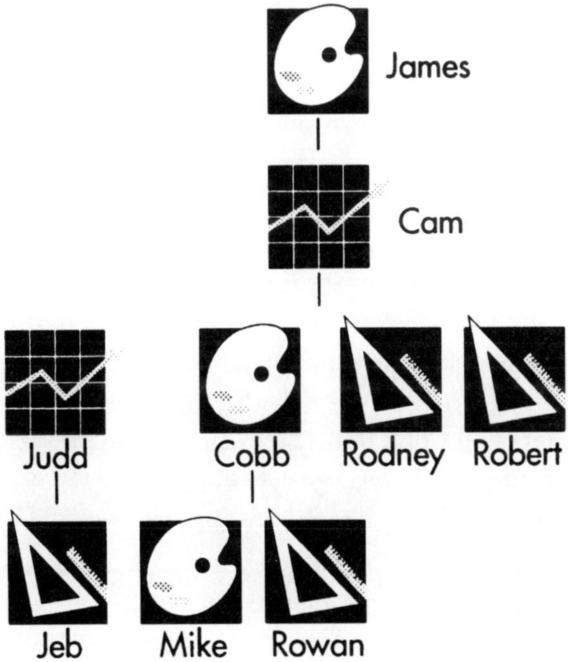

subtly, "We have to put someone in there under him to strengthen the analytic capacity." Although seen to be methodical, that isn't enough. Because he is *knowledgeable,* has long experience in the business, his presence is, temporarily at least, tolerated.

Moving right in the organizational chart, we find Cobb. Imaginative, bold, inspiring, and intuitive, he ties for first place as *exciting.* He leaves no one cold or neutral. Overwhelmingly an Artist, he is nonetheless seen by the majority to share the craft virtues: he's honest, humane, people-oriented, and open-minded. Indeed, it is just this craft capacity wedded to his artistic vision that makes him inspiring in the eyes of his subordinates. There is nothing more gratifying than to be genuinely listened to by someone we consider better than ourselves. One of his immediate subordinates said of him, with deep admiration and affection, "You work for a man like him once in your life." Hard drinker, hard player, ebullient and moody, funny and bold, he flies high, falls low, and withdraws—temporarily. He exudes tension. He makes the Technocrats exceedingly uncomfortable. The other Artists alternately conspire with him or, if necessary, try to end-run him to get their way. The Craftsmen chuckle at his histrionics, laugh at his jokes, and defer to his vision. They would follow him almost anywhere.

Another Artist, Mike, reports to Cobb from Canada. Spiritually fellow-travelers, they sometimes get in each other's way. Dedicated to their own visions, impatient with any procedural delays, neither has much time for approval processes or rules. "There's more than one way to skin a cat" could be their motto. Intuitive and honest, entrepreneurial and generous, daring and dedicated, Mike combines even more strongly the virtues of craft and art. He's visibly calmer than Cobb, but his actions speak otherwise. Like a good thoroughbred, he's classy; "given his head," he takes advantage of it to run free. A Technocrat would say "run wild." He looks after his people, genuinely cares about them. He helps them through difficult personal problems. He is loyal, perhaps to a fault. He is warm, open, laughs readily, smiles regularly. He admits, privately, to recurrent depressive periods, but that is not what the world sees. Not at all arrogant, sometimes naive, he will later be bewildered by events.

Hierarchically Mike's twin, and emotionally his friend, is another Craftsman, Rowan. Amiable, open-minded, and sensible, he makes few enemies. Nevertheless, his days are numbered.

Returning to the level of group CEO, we find two more Craftsmen, Robert and Rodney. Responsible, knowledgeable, trustworthy, reasonable, and open-minded, one is quintessentially a Craftsman and the other is positioned as slightly more technocratic on the map. He is seen to be conservative, stronger than conventional. He's more controlled, meticulous, and detail-oriented. But, honest and straightforward to a fault, he speaks his mind. Whatever resemblance he has to the Technocrat in cognitive domains, like being meticulous and analytical, will not be able to compensate for the Craftsman sides of his managerial style.

This, then, was the team as the final period of my study opened: led now by a Technocrat, still influenced by a previous Artist chairman, seconded by two Craftsmen, an Artist, and a Technocrat. It was an interesting mix. How did they think and what did they do?

Strategy and Style

As noted earlier, the strategy of the ensemble had already, with the elevation of Cam to the number-two spot, been evolving in a more well-defined sense. Gone long since were the bits and pieces James had picked up along the way. Now this precision was to intensify: Anything not directly connected to financial services was divested. The industry recipe, the conventional wisdom, of participating in all subsections of financial services was followed to the letter. That year's management report was testimony to the new concerns. It read:

> Better-informed consumers, and a growing pool of savings, have brought massive change to the financial services industry. To keep up with this revolution necessitates being able to offer the consumer a broad range of products. In order to manage such a system requires, at the same time, huge continuing investments in information technology. The result is that your organization needs scale to defray these costs. The market is irreversibly committed to the integration of services.

Scale was no longer sought in order to accomplish social goals but to reduce costs. This melody began to be played as early as 1983; while James had never been inclined to produce written strategy documents, the elevation of Cam had begun to result in their proliferation. By 1985 the mission statement, conceived centrally and distributed to the divisions, came close but did not use the then-prevailing vocabulary of one-stop shopping. It asserted:

> The group is actively seeking to establish or to join existing distribution networks to reach more consumers. It is planning to develop a national brokerage network to promote the sale of life and general insurance products on an integrated basis. In addition, it is seeking to become associated with retail business firms to test the merchandising of financial services through retail distribution outlets.

Thus the strategy was increasingly clear and well defined. The Craftsmen didn't buy the idea of one-stop shopping for financial services but didn't really care. Interviewed in 1987, one told me, "I just keep my head down and look after business. That's what counts." To him, networking and synergy were theoretical notions and probably, by definition, impracticable. The Artists didn't in any way wish to violate the strategy as stated—unless it got in their way. They largely ignored it, did not pursue the development of similar documents in their own jurisdictions—much to the consternation of the Technocrats—and went about doing what they thought was important: building assets. While the assets of the Technocrat's sector remained steady, the artistically controlled sector ballooned. Quintupling in Canada, doubling in the United States, overall the sector trebled.

Developments

Although London-based, Cobb was almost never at the head office. Spending half of his time on the Concorde, he was regularly in Canada or the United States, called upon to be part of acquisition negotiations. When he was home, he was busy with his own deals. In 1987 a broker approached him with the rumor that a certain retail chain in the United Kingdom would be open to a proposition to join forces with a financial services company. Cobb recommended a second strategic alliance. This became the fifth major axis of development, paralleling the Sears experiment in the United States, and someone had to run it. Too busy with offshore and domestic developments, Cobb was persuaded that it should be supervised by someone else. One of his vice presidents, Bill, was promoted into the slot. Bill was a Craftsman with artistic leanings. Honest, humane, never distant, uncompromising, or detail-oriented, he was sometimes seen to be funny, emotional, and entrepreneurial. He was to report through the Technocrat, Judd, to the chairman; this because, they said, Judd was based at the head office and had more time. Bill was less than pleased to report to Judd—he knew him—but he wanted the challenge.

Centralization/Decentralization

Although the ethic and the official language of the group had been decentralist—let the divisions and their independent boards run their own show—there were forces at work moving in the opposite direction. First of all, there was sheer size. As the group grew and diversified, both nationally and internationally, the center felt cut off from the action yet globally responsible for it. Second, the strategy itself, of participating in all subsectors of the financial services industry, seemed to some to carry with it the necessity of synergy; that is, if you sell general insurance to a client, you may also sell life insurance or a mutual fund. Thus, the various units must be capable of cooperating, and this was a delicate and complex question involving, among other things, complicated systems of sales remuneration. Third, there was a power struggle: Would the center be most important, or would divisional management and their independent boards? This struggle, not at all unique to this particular group, was fed by personal jealousies and by fundamental stylistic differences between the players. Thus, an Artist said of Cameron, "He always talked decentralization. That was what was *expected.* But his actions were always centralizing." To a Technocrat, with his deep-seated need for orderliness and control, decentralization spelled anarchy, and this he could not abide. Given to intellectualization as a primary defense mechanism, he found in the first two factors—size and synergy—the perfect justification for his actions. All these issues were to come to a head around technology.

Servicing an increasingly diverse and demanding clientele meant massive investments in information technology. Synergy, it was argued, necessitated that those investments be harmonized, that all systems be made compatible. Separate systems development would involve huge outlays, possible duplication, higher than necessary operating costs, and potential frustration of the synergy objective. Freed from the daily management of his own sector by the presence of a dependable Craftsman and physically lodged at the head office, Judd found it logical to assume responsibility for technological harmonization. It was also logical that the divisions would fight Judd's assumption. The Craftsmen and Artists running those divisions saw his move very clearly as an incursion into their operating authority, as a restriction on their capacity to innovate in an area vital to competitive advantage, and as a patent power play.

They also believed that the future of information systems was toward distributed processing. Who was right? In any case, this was not really a contest of facts. In a 1987 interview, one Craftsman offered an explanation: "[Judd] doesn't care about costs or about synergy. He's simply trying to increase his empire and his power." Backed 100 percent by Cameron, Judd was to succeed in divesting the divisions of important areas of discretion.

What was true for technology was also true, but to a lesser extent and with much smaller financial and discretionary implications, for marketing. Synergy required tapping into the client databases of each subsidiary and developing common marketing approaches. In this case, a VP at the head office, reporting to Judd, played an encouraging role, not a dictatorial one. Nonetheless, any hesitancy or apparent lack of enthusiasm on the part of divisional management did not go unnoticed.

Synergy meant that plans were needed, indeed, insisted upon. A head-office executive, once again under Judd's supervision, was given responsibility to "work with the subsidiaries" to develop written strategic five-year plans that could be "harmonized." All but one division, run by a recalcitrant Artist, cooperated; of strategic plans he said:

> They wanted five-year plans, but that stuff's just an excuse not to work. All they do is produce tons of paper that just goes into a shredder eventually, anyway, so what's the point? Can these guys produce and sell a product? No, they say in their plans, "We will produce and sell such and such a product," but they can't do it.

He would later pay for his intransigence. Once a year, a two-day planning retreat was organized, not to consult the divisions but to make sure everyone was "on side" with the head office.

All this planning activity spurred an intensification of divestitures. If it didn't exactly correspond to plan, an activity was divested. There was no time to waste on watching dreams grow up. Thus, in this period, assets accumulated in one sector were more than compensated by divestitures in another, such that, overall, asset growth remained almost flat. This divestiture activity was steered by the head office and by—guess who—Judd.

Gone now were the days when the annual reports talked of philosophy, humanity, love, and friendship. With increasing regularity, new words began to replace them. Instead of *friendship*, it was *efficiency;* instead of *love*, it was *rationalization* and *economies of scale* (emphases are mine):

> In our search for greater *efficiency*, the collaboration of our various companies has brought great benefit. . . . Putting finance and human resources functions together allows us to benefit both from scale economies and from the possibility of the latest technological advances.
>
> From the point of view of *rationalization*, we have proceeded with the reorganization of [company X], with the sale of [company Y] and we continue to concentrate on *coordinating* technology and marketing.

Coordinating indeed. By now, the areas of finance, human resources, information systems development and data processing, planning, and marketing were all being "coordinated." *Coordination* was a euphemism for control, and *control* was a euphemism for centralization. Its byproduct was demoralization.

Management Style

The need now to have formal plan approval from the center gave scope for more intimate inquiry into the managerial style in the operating divisions. One Craftsman put in his mission statement that his organization should be, among other things, a place to have fun. He added his intention to create the position VP of the Impossible. Cam was far from amused. Both were removed from the mission statement. In another sector, a Craftsman decided on a regular schedule of visits to all his sales outlets—management by walking around. Cam considered this a complete waste of time; in the eyes of the chairman this man had more important things to do than to go around talking with staff. One of those more important things was to produce profit.

The pace of acquisition activity, although exciting on the street and stimulating for the price of the stock, was not without its downside. It took time, effort, and a great deal of money to weld the various units into workable teams—always more time, more effort, and more money than forecast. And acquisitions were not always successful. One in Canada was proving disastrous, draining the surpluses normally generated from that region. Convinced that the stock price would follow quarterly earnings and not growth prospects, Cam and his spiritual sidekick Judd began a campaign to boost short-term profits. The emphasis in planning documents began to be placed not on long-term development but on return on investment (ROI) targets. What began as an exercise in harmonization of visions ended in marching orders: Produce 16 percent ROI or get out. A Craftsman insisted, with disdain, "Sixteen percent ROI by 1995 is a joke; we'll be dead by 1995."

It wasn't in the script, but it was in the cards.

7

Climax: The Triumph of Technocratic Illusions

For considerable periods the four oboe players had nothing to do. The number should be reduced and the work spread more evenly over the whole of the concert, thus eliminating peaks of activity. All the 12 violins were playing identical notes; this seems unnecessary duplication. The staff of this section should be drastically cut. . . . No useful purpose is served by repeating on the horns a passage which has already been handled by the strings. It is estimated that if all redundant passages were eliminated the whole concert time of 2 hours could be reduced to 20 minutes and there would be no need for an interval.

Alleged work-study engineering report, 1950s[1]

Cam was by now beginning to feel that he had too much on his plate; in particular, the regulatory authorities in all the jurisdictions were very active and required constant watching. As the visible representative of the corporation, the chairman not surprisingly found this responsibility falling largely on his shoulders. In addition, he was feeling his age. Anxious to provide for an eventual smooth succession, he felt it was time to move one of the managing directors into the number-two spot. His choice lay between an Artist, Cobb, running by far the largest operation, and the Technocrat, his head-office sidekick, Judd. It came as no surprise to anyone that Cam chose Judd. "They're birds of a feather; they flock together," commented one Craftsman. "The handwriting has been on the wall for some time," added another disillusioned Craftsman. Thus, midway through this last period, the organizational structure looked like the one below.

Profits: The New "Strategy"

With two Technocrats at the top, the interest in synergy curiously began to wane. It was time to get serious. Maybe this synergy business was a distraction; the thing to concentrate on was running each core business to the best of everyone's ability. A Technocrat described it in the following terms:

Structure late 1980s

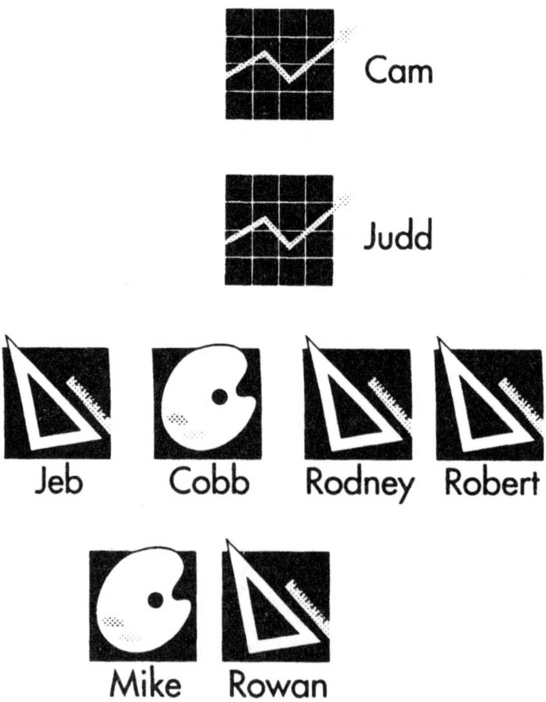

> In James's day, it was a cult of personality. Things happened without him realizing it. In the sense that it was all *opportunistic*. There was no strategy. Why were they in Canada, the U.S.? It all lacked *logic*. Then they brought in the pros, Cam and Judd, to put some *order* into it all. They built the group around a concept that does not work—synergy. One-stop shopping, like a drugstore. But, there has been no revolution, just evolution. Everybody is just getting better at what they do in individual sectors.

But synergy had been what everyone said was necessary; it was the conventional wisdom of the day. But a new era had begun, the era that said "stick to your knitting," so now the Technocrats turned to it. The Craftsmen, who had always stuck to their knitting, were amused by this new turn of phrase, but not amused by what would then be erected in its name. Getting better meant getting more profitable. Right now. Today.

The Canadian Scene

Once contributing substantial profits to the group as a whole, Canadian operations were now a drain on cash flow. A two-year-old acquisition had gone badly awry. The due diligence at the time of acquisition had been badly flawed, and the underlying assets turned out to be in much greater disrepair than imagined. It was hemorrhaging cash. Mike, the Artist in charge there, was under considerable heat, as was his London boss, Cobb, since Canadian operations were consolidated into Cobb's sector's earnings statements. Mike kept asking for time; Cobb was blamed for giving it to him. Formally now Cobb's boss, Judd insisted that Cobb place someone in charge "over there" who knew what he was doing. To Judd, that meant a technician, an actuary, an accountant, a finance guy. Mike was temporarily kicked upstairs, removed from day-to-day operating authority and, within six months, fired.

> COBB: I shielded Mike for a long time, but I couldn't do it indefinitely. He was resisting. Talking back too much. He

always said what he thought and that was very dangerous. They never had patience with investments. They panic. At that time, I couldn't stave off their assault. My own power base was too weak.

MIKE: It wasn't Cobb's fault. It was those guys'. They have a total disrespect for skills, experience. Even when I was a kid I sought out older people to listen to. They think whiz kids can run anything. There's no tolerance for error. Trial and error is how you learn. If they had let me solve that hemorrhage I could have stopped loss. But no, they had decided that I created the problem so I couldn't solve it. It was the excuse they needed. They were just waiting for me to fall. I should have seen it but I didn't. I just couldn't believe they could be so stupid. I used to pray before meetings, "Please, God, help me to keep my mouth shut," but it never worked. They wanted five-year plans, but that stuff's just an excuse not to work. All they do is produce tons of paper that just goes into a shredder eventually, anyway, so what's the point? Can these guys produce and sell a product? No, they say in their plans, "We will produce and sell such and such a product," but they can't do it. They don't know salesmen. They can't stand them. Me, I honor them. I treasured what they did. You have to love what you do to do it well. They don't love anything. Me, I went on my instincts. Ninety-nine percent of the time it worked; I made one mistake. I'm out. But what will they do now? They've picked all the fruit from the tree to artificially inflate earnings. What will they do now? There's no fresh air. They're like the Russian technocrats, destroyers. That's why Russia's crumbling. Parasites living off the productivity of others and ultimately destroying that which they live off. . . . Okay, I screwed up, but do they ever admit they've done anything wrong? For them, people have to be perfect, all assets, no liabilities. It's so unrealistic.

In this cri de coeur we hear the Artist's disdain for planning exercises. He paid for that. We hear the belief in instincts. He paid for that, too. We hear once again words we heard from James—love, honor, treasure.

> CAM: You know it's the only thing I regret; I should never have let him buy that company. I had strong doubts, but he was so insistent. It's too bad.

Cam had been the chairman of Mike's board. Ultimately the acquisition was his responsibility, but he accepts none. What we don't hear from Cam is more important than what we hear. Cam is not remorseful about the firing. Mike made a mistake and, as surely as night follows day, had to be fired for it. He regrets, instead, not having maintained a fail-safe system, not having *controlled* the situation.

> A BOARD MEMBER: They should never have fired Mike, but he was just like James and Cobb. Generate new ideas and get on with it. Entrepreneurial personalities, impulsive, opportunistic. But Cam was not . . . I don't know, maybe it's a phase. You know, growth, passion, followed by a period of reorganization ad nauseam, then a new growth phase. But the only way growth will happen is if there is a hostile takeover; it's not these guys who will create new opportunities.

Mike was replaced by a Technocrat.

The United States

The acquisition that had, in one coup, doubled U.S. assets had also doubled Rowan's headaches. Head office had to be moved into larger quarters. Sales teams' premerger compensation packages were completely different. Accounting systems were not up to handling the extra load. Things got lost. Computer systems were incompatible and too small. There was no control, or at least seemed not to be. Profits were adequate but below expectations. Postmerger economies took their jolly good time materializing. In short, a mess. Rowan was a Craftsman. He believed these things would take time but would eventually sort themselves out. He didn't feel he could push the system any harder than it was already being pushed; that could create worse problems—discontented sales staff, diminished production, morale problems. These things simply take time. It's *unrealistic* to think otherwise.

> ROWAN: I don't know why it happened. Profits were on budget; bonuses were paid. I guess they just wanted more. The spotlight began to shine on us when things started to go badly over there. I guess they needed us to compensate for it. They wanted everything faster and more, but there was no time. They didn't really understand our situation over here. Too distant from the coal face to understand how the business works. Gave them exaggerated expectations. In person, everything was sweetness and light. I suppose they're honest, but they're certainly not straightforward. Now the whole group is like an individual with a frontal lobotomy. No memory. They've sacked anyone who knows anything about the business. They don't respect experience. They hire based on type, and type is somebody who will do the work without raising any objections, any implied criticism. They have a clear view of management style and anything else is aberrant. Nobody who wants to make a mark will ever succeed with them. Their strategy was very murky. They seemed only to be fascinated by tinkering with the thing—structures, systems. Now they've used up all the goodies, the organization is really anorexic. They seemed to think that you could grow and expand and not have an impact on the bottom line. But any schoolboy knows better. They're like the little old lady who says, "I want capital gains and income." Totally unrealistic.

We hear again the Craftsman's bewilderment. How could they think that the system of people could be pushed any harder? Those who value straightforwardness, honesty, and realism find Technocrats unreasonable, unrealistic, even childish in their managerial theories, adults with a child's dogmatism.

> JUDD on ROWAN: He simply was not a manager. He was over his head. He was an adviser. He was volatile, unpredictable, imagi-

> native, visionary. A nice guy, very amiable. The kind of guy you might seek ideas from but not the guy to run something. Didn't know how to work.

Judd gives us a very typical Technocrat's response. To him, Rowan was a very likable guy. He had nothing personal against him. There is no affect, no emotion, involved here; it's simply a matter of judgment and being realistic. Rowan couldn't do the job; he had to go. Nothing personal. No hard feelings. Note, as well, his suggestion that Rowan was not a hard worker, a suggestion contradicted by others closer to the scene. Judd expects a managing director to give 150 percent of his or her time to work. Leisure is a luxury.

> A BOARD MEMBER on ROWAN: He was very well mannered and considerate. He fired people but did it nicely. He was hardworking and very dedicated. Very open-minded. Sometimes ahead of his time. I don't know what they wanted.

Rowan was replaced by a Technocrat.

The United Kingdom

The profit and loss statements of Canada and the United States were consolidated directly into Cobb's financial statements. The stock market crash of 1987 was having a negative impact on his earnings, even though the losses could be amortized over several years. This made him more vulnerable. In addition, Cobb's handling of the U.S. and Canadian situations, or, what was more the prevailing view, his failure to handle them, confirmed opinions about his "managerial capacity." "He's just not a manager," they said. It had not helped, of course, that he was less than enthusiastic about "harmonizing" the information technologies, or marketing, or synergies. And it had not helped that he was the only other serious contender for the top spot. It was decided that he needed "assistance," a more "professional" person under him. This pro was to have a seat on his board, become the managing director, and Cobb was to be kicked upstairs, where his talents could be put to "better use." Cobb had seen this pattern before. He knew very well what it meant. Ross, another Technocrat, was to take over.

> A BOARD MEMBER: As soon as Judd won, Cobb lost heart. It was, of course, compounded by the arrival of Ross and his perpetual past run-ins with Cam, but basically the die was cast. He had to go. There was no way to stop the momentum. There was no way he could get along with them.

Cobb's own board didn't like it—there was a minor and futile revolt—but as another board member put it, "What the shareholder wants, the shareholder gets." Judd and his man Ross, a Technocrat, were in. Cobb was out.

In general, the Technocrats liked Cobb, or so they said. They call him *intuitive, imaginative, visionary*—in short, all the artistic features that make up his personality. He just couldn't do the job.

> A TECHNOCRAT: He was *Star Trek*y. He produced a hundred new ideas but only one of them would be useful. It's terribly inefficient and ineffective. The people underneath get frustrated. It's awe-inspiring but eventually demotivating. Our world cannot do without them but our organizations can.
>
> A CRAFTSMAN identifies the nature of the "mistake" COBB made: He was the visionary, spent his time on acquisitions and foreign operations and wrongly trusted his staff to run the show at home. But he never implemented the head-office system that would have allowed the decentralized management of the subsidiaries. So the wheels started falling off all the carts.
>
> COBB: It was all about strategy. What is strategy anyway? Grand plan? No. You try to instill a vision you have and get people to buy in. The strategy comes from astrology, quirks, dreams, love affairs, science fiction, perception of society, some

> madness probably, ability to guess. It's clear but fluid. Action brings precision. Very vague but becomes clear in the act of transformation. Creation is the storm. They always built things around "barring unforeseen circumstances," but that doesn't leave room for the unforeseen. It implies you know the future. It's a straightjacket, narrow walls. The unforeseen is not supposed to exist, but you have unforeseen government policy, economic change, fraud, loss of key people, technological change. These people live outside of reality. You should never allow the plumber to be the engineer, because he'll run the pipes across the top of your machines because it will be easier and cheaper to repair later. Of course, in the meantime, people can't work.

Around the same time, two Craftsmen, Rodney and Robert, also lost their jobs. The Technocrats said things like: "They just couldn't do the job"; "He didn't evolve with the times"; "He was too old-fashioned"; "He wasn't tough enough"; "He was boring"; "He has no personality." They were replaced by David, a Technocrat, and by George, a Craftsman (a mistake they would later correct). The latter was to find himself exceedingly uncomfortable in the days to come.

> A CRAFTSMAN (with an MBA): Their kind strip the essence out of organizations, empty them of their sense. I think it's a product of the engineers and MBAs of the 80s, the technocratic mentality. But everywhere it's proved a failure. People are beginning to see it.

During this period, the tenor of public statements changed, as well. The preoccupation with reorganization, rationalization, consolidation, structures, and profit was intensified:

> We are proceeding with the reinforcement [read: firing] of our management teams, with the rebalancing of our activities, and an intense effort of rationalization [read: cutting] of our operations. We are also proceeding with a revision of our operating structures [read: new

> organization charts] in order to promote the achievement of our objectives and improve our performance [read: quarterly profits].

The next year, the same themes were played:

> The progress observed in our results is the fruit of several changes: the reinforcement of our management teams, the rationalization of operations and the concentration of activities.

A New Hand on the Helm

By the end of the 1985 to 1990 period, Cameron had vacated the chair in favor of Judd, over the objections of significant minority shareholders on his board.

> A BOARD MEMBER: We tried everything we could to stop it, but he wouldn't listen. He's stubborn. He just didn't want to hear and, short of a proxy war, there was nothing to be done. Besides, there was no longer any other potential candidate inside. We'll just have to wait. Eventually they'll make a mistake. Things will continue to go badly and we'll be there to pick up the pieces.
>
> ANOTHER BOARD MEMBER: I went to him at the time because I felt sure he was going to choose Judd. I was not necessarily in favor of Cobb; I wasn't sure that he was the right man either. But at the worst, we should recruit outside. I didn't trust Judd; he always seemed to say the right thing at the right time and that made me very uneasy. I didn't think he had the intelligence or the depth to run an organization of this size and complexity. Cam was very defensive, even angry. He said that Judd was brilliant, that he had turned around his sector, and that was that. I was ushered out of his office.

I asked Cam if Judd had ever disagreed with him about anything. "No," he said, "but what has that got to do with anything?" (The Technocrat is submissive and cooperative *until* he gets into power; this is symptomatic of the compulsive personality, remember? Relations organized around dominance and submission.)

With Judd's elevation, the organization looked like the following chart. The Artists—James, Mike, and Cobb—were all gone. So were the Craftsmen—Rowan, Rodney, and Robert. Two other Craftsmen, George and Bill, were demoted; a third, Jeb, knew he was next. He was to be lent assistance by a new recruit—to all appearances, another Technocrat. In fact, in the closing days of the study, Jeb was kicked upstairs and was on his way out. (He would be fired a year later.) In the end, only Technocrats had any power or influence.

Strategy

A report concluded:

> Mirroring a worldwide trend among large financial groups, we initiated in 1989 and continued in 1990 an extensive program under

Structure 1990

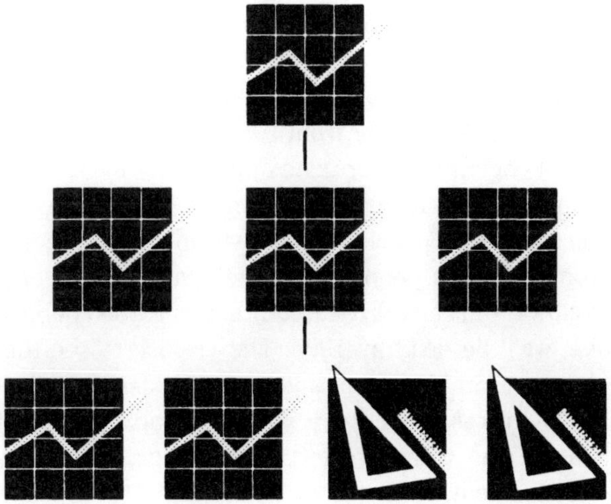

> which operations were regrouped, assets sold and activities rationalized. As a result, we now operate in two sectors: insurance and banking. . . . New chief executives have been appointed and our strategy is profitability.

"Our strategy is profitability"—many felt this formulation was a contradiction in terms. Profitability is an outcome, not a strategy. Asked what the current strategy was, one remaining Craftsman replied, "I honestly don't know." A relatively new Technocrat said, "I wish I knew. They seem to concentrate on short-term profits. I guess that is a strategy. There's a meeting coming up. Maybe then I'll know more." He didn't. But that doesn't matter because he has since been fired. Another Technocrat, closer to the center and part of the action, adds, "Eventually? Alliances. *Everybody knows* [read: you imbecile] you have to create alliances. In the meantime, we've had to create profit." The new buzzword is *strategic alliances;* it is on all their lips.

> A CRAFTSMAN: They don't know. Billions have been spent on strategic planning, but that's not the issue. It's strategic thinking. The plan comes later. It fits into the "profile" of what you want to be when you grow up. James and Cobb had the "profile," the vision. Strategic planning is the death knell of strategic thinking. Once it's on paper, the job's done! Even if they had a vision, how would they get it done? There's no managerial continuity. At this year's planning meeting, there were four out of fourteen people left over from 1988. Every two years there's a new chief executive. There's no opportunity to fail, so there's no continuity. They focus directly on profit, but they'll never get it because profit comes from the vision and the people, and they won't invest in people. If you look after the people, the profit follows. You can't drive at it directly. Sixteen percent ROI is a joke; we'll be dead by 1995. They refuse to see this. You can't correct a problem unless you see it exists. It's like me. I look in the mirror and I see a young fullback, not a balding, middle-aged man with his chest on his belly. You have to see reality to change it.

The New "Team"

The Technocrats view each other:

> He's a lightweight. They all are. Amateurs.
>
> People are too hard on him. He's not a lightweight. It's true he's not brilliant, but he's put order into the shop.
>
> He's an ass.

Not much team spirit there!

Technocrats on previous eras:

> In James's day it was easy. Things moved slowly and you could afford to make mistakes. Now when you make a mistake you know it instantly and the consequences are much bigger. You just can't take those risks now.

Technocrats always feel that the past was simple and that their times are more complex and difficult and demand their much greater sophistication. They are narcissistic.

> James, Cobb, Rowan, they were all dreamers, not entrepreneurs, not builders. They're what I call advisers. Now we have builders. They're conservative, of course, but if you're not conservative you'll lose everything. James was always talking in metaphors. It took a professional like Cam to put some order into things.

Well, the organization started full of optimism and reached out to the world. It grew and consolidated by art and by craft. Now, it had lots of *order* and was full of *professionals,* all watching the systems and the bottom line. But the Craftsman who told me in 1990, "Sixteen percent ROI by 1995 is a joke; we'll be dead by 1995," was wrong. As I sit here at my computer today, in February

1994, James's dream is already stone-cold dead. The vision, and the organization it spawned and nurtured, collapsed. With poor profit performance and stock trading at less than half book value, the company was easy prey for a takeover. Some people escaped from the wreckage without a scratch. But there were many casualties, and some lives were damaged irreparably.

Let's examine why.

8

Dénouement: The Collapse of Vision

We are as ambivalent about heroism as we are about the workaday goals that it sacrifices. We struggle to hold on to a vision of the incomparably higher, while being true to the central modern insights about the value of ordinary life. We sympathize with both the hero and the anti-hero; and we dream of a world in which one could be in the same act both.

Charles Taylor, *The Sources of the Self*[1]

What does this 15-year story tell us? I'm sure you have your own ideas, and that is as it should be; everyone hears different music in the same recording. Here's what I hear.

I think there are a lot of reasons for the collapse. Some argue that there's been a big shakeout in financial institutions—U.S. savings and loans institutions and Canadian trust companies and insurers, for example. Well, that's true, but why was this particular group shaken out? What made it and not another more subject to, more vulnerable to, these forces?

Another argument often proposed is that the group was *overextended*, that it got carried away by acquisition euphoria and had to pay the piper. Well, okay, but let's look closely at this argument. It's certainly true that first James, and then Cobb and Mike with James's moral support, believed that the primary goal was to build a strong enough asset base to be able to compete globally. Cobb argues that you have to shoot high to attain your goals, because there is inevitable slippage from them. On the back of an envelope, he drew the diagram below for me.

Here's how he explained it:

> Say you have a goal for the year 2000. You have to shoot at a target, point B, much higher than the goal, point A, in order even to come close, because the world is always changing—competition, obsolescence, government, deaths. So, you acquire, you do R&D, you look for new markets. If you drive directly at the goal A, you will inevitably fall short and arrive at point C. The cost-cutters think they're driving at the goal, but all the while they've got their heads down looking at the books, the whole world changes around them, and the whole thing collapses. At 1990, the two approaches may look the same, but by the year 2000, it looks different.

This is the Artists' explanation for the collapse. Too much time and energy had been devoted to structures, to reorganizations, to systems, to short-term profits, and too little time devoted to development of new products and new markets.

You might reply that this was inevitable. The group was big—$10 billion in assets—and these centralized systems *had* to be

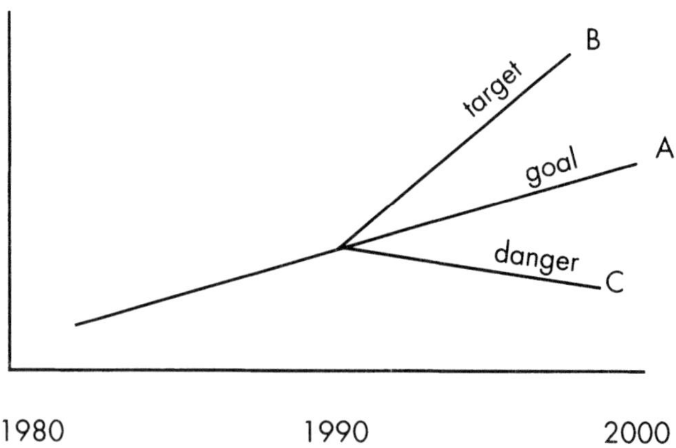

developed. But why? There was no profitability problem in James's day. James didn't need elaborate centralized control mechanisms, because he left the discretion in other people's hands. He let the CEOs run their businesses. He sat on their boards and sometimes asked embarrassing questions, but basically, the various divisions ran their own shows. He believed they were better at it, knew more about their business than he did. He trusted them. Furthermore, he had other castles to build. The Artist *always* has new ideas, new projects; it's his nature, his *character.*

Yes, but *synergy,* you say. Synergy necessitated making them all work together. Well, so it was thought at the time. It was certainly the conventional wisdom. But within three years synergy was judged no longer necessary, so I have a hard time believing that it was logical, a rule carved in stone. Carved in papier-mâché, maybe. Even the Technocrats abandoned the rationale once they'd used it to achieve their centralizing goals. They *need* to centralize, do you see? They will use any intellectualization, any handy excuse, to get it. They're *frightened.* They're afraid of failing. They believe that only *they* possess *Truth.* Only *they* know how to do things. Only *they* are competent. This is their nature, their *character.*

The Craftsmen agree with this diagnosis, say the group collapsed because there was too much "tinkering with the thing." But one Craftsman adds another, and in his mind more vital, reason. "Now the whole group is like an individual with a frontal lobotomy. No memory. They've sacked anyone who knows anything about the business." The Craftsmen believe you can't run something unless you have experience. You will inevitably make clumsy, unnecessary mistakes. You will, by this clumsiness, by this lack of skill and dexterity, alienate and undermine traditional markets. You can't stick to your knitting if you don't know what knitting needles look like.

The Craftsmen also believe that, even if you have a good strategy, you have to work through and with experienced people to get the job done. They think, therefore, that continuity of personnel is the sine qua non of bringing the vision to fruition. One told us:

> Even if they had a vision [which they don't], how would they get it done? At last year's planning meeting, there were only four out of fourteen people left over from 1988. Every two years there's a new chief executive.

New people brought on board continually have to learn the business from scratch—particularly, as was usually the case, when they don't come from the industry. "They think whiz-kids can run anything."

But why is there no continuity? Because Technocrats can't allow mistakes. Mike said, "Okay, I screwed up, but do they ever admit they've done anything wrong? For them, people have to be perfect, all assets, no liabilities. It's so unrealistic." If you ask a Technocrat, they'll tell you this is not true. They know, better than most, all the management literature. They're the ones who read it. They know all about "empowerment," all about "loyalty," all about "participative management." They know, theoretically, that you have to allow people to make mistakes. They can give you learned discourses on these subjects. Judd gave me a long lecture one day on "total quality management" and how it was the new secret to success, to "empower" people. This was, it goes without saying, after he had gotten rid of all the "dreamers" and the "incompetents" and replaced them with "professionals."

In Chapter 3, I told you that walking into this organization was like walking into a morgue. Well, in one way it was, but not in another—there were no cadavers hanging around the halls. Actually, nobody hung around the halls. Nonetheless, the absent—the people who left or were fired—made themselves felt. There was an atmosphere of tension and fear. "Who will be next?" "Will it be me?" "Am I 'professional' enough?" One Craftsman told me, "I'm trying to turn myself into a robot, but I'm not succeeding very well." As a result, people worked to the rule; that is, they did nothing that might later be construed as a mistake. They stopped taking any risks. And they looked for other jobs. Those who were not fired left voluntarily. There was a hemorrhage of experience. Although Craftsmen are not brilliant, they see this hemorrhage and its implications very clearly. They're wise. That is their nature, their *character.*

In my book, the vision collapsed because there were no Artists left, no new "bridges thrown out toward an unseen shore"; all the shores were too well defined. According to the Technocrats, the Artists weren't to be trusted; they were fools and dreamers. And there were no Craftsmen left to engineer and build the bridges. Craftsmen were caught in the cross fire; insufficiently glamorous and visionary on the one hand, and insufficiently rigorous on the other. They couldn't win against the technocratic juggernaut. So all that was left was just a bunch of technicians, virtuosos, paint-by-numbers managers. But management is not paint-by-numbers art. If it were, we wouldn't be in such confusion today.

The Technocratic Triumph

But how did they do it, these Technocrats? Why were they so powerful? How could they succeed in getting rid of everybody? One, they're brilliant. They really are. Minds like steel traps. They come into board meetings with their five-part plans and strategies and projections, with their slogans, their grandiloquence and their recipes, and it's pretty impressive stuff. It's intimidating to ordinary mortals. By hijacking the language of vision, by imitating it, they seem brilliant. Apart from their technical skill and ingenuity, their virtuosity, these *grands parleurs* (big talkers) have something else: They're *strategic about human relationships.*

Whether by Machiavellian intention or by the necessity of their characters, they think very long-term about human relationships.[2] They undermine others gradually. They start by making fun of someone or of their ideas:

> (*Friendly chuckle*) James? Well, that's James (*chuckle, wink*). He cares about those kinds of things. You know, emotional, sentimental. But don't worry, we follow up behind him and look after the serious stuff.

But of course, you can't get rid of someone who has the status of folk hero, so you maneuver them into leaving of their own accord. After you have insidiously undermined their credibility, first by implying that they're not to be taken seriously and then by indicating that modern management demands something more sophisticated (by the way, did you know that sophisticated originally meant "spoil the simplicity or purity or naturalness of"?[3]), you gradually remove them from power. Judd and Cam, ever so gradually, removed James from all the important operating boards and committees. I sat across from James one day at a luncheon during which a list of new appointments was announced. His name was no longer on the list. He leaned over to me and whispered, "They could have told me before the meeting, don't you think? Just as a courtesy?" But no. He might have kicked up a fuss. They might have had to defend themselves publicly. James had literally made these people's careers. He built the empire; he hired them; he handed them power and they used it to stab him in the back. They pushed him and pushed him to the limits, and finally he had no choice but to resign. With him gone the rest was easy.

Isolation is another tactic Technocrats use to great advantage. You pick them off one guy at a time. No mass firings, just sniper fire. First, you undermine someone's credibility by mocking it; second, you imply that he needs (poor fellow) help; then you install "help" under him; then you kick him upstairs where his talents can be put to "better use"; then, what do you know, "It's not working out. He's going to have to go"; finally you pay him a big separation settlement and he's gone and you move on to the next one. After Mike was fired, I said to Rowan, "Watch your back. You're next." He said, "What do you mean?" You see, he just couldn't believe it. It was like a nightmare, and he was sure he'd wake up and people would have come to their senses. It's an old rule, no? Divide and conquer.

But the board! you say. Where was the main board of directors? Surely it could see the pattern. Some members did, but they were in a distinct minority, especially since Judd and Cam had increasingly loaded the board with friends and doubled its numbers. More important, however, each incident seemed isolated, minor, justified. The new members couldn't see the pattern because they hadn't been around long enough to see patterns in anything. They didn't know the executives who were being rounded up and corralled. Even the old-timers didn't say anything publicly. One, who said "They should never have fired Mike," knew that he himself would be gone the next day if he spoke up. He was still clinging to the faint hope of being able to use his influence to minimize the damage. It's fine, in retrospect, to criticize him, but at the time his motives and his intelligence were both beyond reproach. He didn't have any power. No one else did. The decks were all stacked. The dice were all loaded. The game was crooked.

About That Nasty Word, *Power*

What really happened here was one colossal mistake, and James was the one who made it. He knows this now. "It's all my fault. I'm the one that put Cam in power." The thing about Artists and Craftsmen is that they admit to their mistakes. That's life. That's what humans do. They make mistakes. Sometimes small and sometimes big. But to James's everlasting regret—he's heartsick about what happened—this was a big one. The selection of one wrong person led to the destruction of everything he built.

James, an Artist, chose his opposite to succeed him. He was naive about Cam. Remember his response to my question at the time? "Could the organization be taken over by bureaucrats?" "No," he said. "Its entrepreneurial character is too well entrenched. It's too strongly ingrained in all the people." Apparently not in all the people. James got fooled, but he got fooled for very good reason. The Technocrat is submissive, even obsequious, with bosses, but authoritarian with subordinates. The Technocrat is a chameleon, and changes color depending on the surroundings. In the early 80s, the Technocrats trumpeted their entrepreneurship; it was then conventional wisdom. They were surrounded by entrepreneurs. In the mid-80s, they trumpeted their managerial sophistication with "synergy" and "one-stop shopping" for financial services. It was conventional wisdom; all the "experts" said it was the right thing to do. In the late 80s, the trend was to "rigor," the "bottom line," "stick to your knitting," so that's what they did. They competed for how many people they could lay off. Today they're preaching "reengineering," "total quality management," "empowerment," and "strategic alliances." This last they particularly like because to them it means getting in bed with a big, strong partner who can protect them from a very scary world.

James never *talked* about strategic alliances. He didn't even know the term because it hadn't yet been invented when, 30 years ago, he created one with EFC. He did deals; he didn't *talk* about doing them. He didn't talk about trusting people or about decentralization, either. He just *did* it. It was second nature to him. And all the Craftsmen—Rodney and Robert and Jeb and the others—they didn't *talk* about "empowerment" or "total quality management" or "learning organizations" or "mentoring" or any other fancy recipes. They just *did* it. It was second nature to them. As Osborne told us in an earlier chapter, craftsmanship is all about a *cult of excellence.* You don't have to tell Craftsmen to be concerned about the quality of their product or service. It comes as easily to them as breathing. It's how they live life. In fact, maybe it's no accident that all sorts of recipes for quality, learning, empowerment, participation, teamwork, mentoring, and so on have proliferated during a time of universal technocratic triumph; we have had to reinvent what the craftsman assures naturally.

In any case, Cam didn't repeat James's mistake; if he had appointed his opposite to succeed him, some equilibrium might have been restored. But no. Technocrats, when they value anyone at all, only value other Technocrats. Only Technocrats are competent. Thus, Cam appointed Judd, and the two together chased out all the others and installed five more *competent* clones—or, let us say, quasi-clones, because a couple of them were quite emotional and volatile (of the blast furnace variety). The organization tipped fatally toward disequilibrium, like a boat that takes on water when everyone leans to one side.

This happened because power cannot be distributed equally. It may not be politically correct to say so, but some people don't deserve power. They're dangerous. They abuse it, and not just obvious sickos like Hitler. Some basically moral people are blind. They can't help it. And they can't be taught because they live in the closed circle of their own logic. Competence, for example, is serious; it's not funny and it's certainly not amiable. They can be taught to say words like *empowerment,* and even to believe them, but they can't walk the talk. The intellectual graft doesn't *take* on the affective tree.

Does all this mean that Technocrats should be marched out and shot at dawn? Of course not. We are all as much to blame as they are. We let them take power. Why? Because we're scared, too, and their formulas are very reassuring to us. They seem to know the way to the future. They have a three-part response to every question. We can sink into a comfortable dependency safe in the knowledge that they will look after us, secure our futures. *But it ain't so.* They can't.

And we give them power for all the reasons that philosopher Charles Taylor talks about, although he says it so much better than I could ever hope to do. The social and intellectual reasons. We, as much as the Technocrats, distrust emotion and passion. Passion, of a religious variety, caused the inquisition and certain other, more modern, unspeakable horrors. We distrust dreamers; dreamers can turn out to be psychopaths.

And tradition, we distrust it, too; clinging to it kept corrupt absolute monarchy in place, and we had to get rid of that. We have lost our respect for craft, except in the flea-market context, because to us it also seems too slow. We want answers and solutions right now, today. There's no time to waste. Have you noticed how everyone says things like, "It's an age of discontinuity," or we are living through a "rupture" or a "crisis," meaning, "Hurry up! Do something!!!" This, too, is inimical to craftsmanship. Every age imagines itself faced with unprecedented rapid changes and turbulent times, and it's just narcissism to imagine that our age is somehow exceptional. Only history can make a relative judgment. In France, a fourteenth-century man stepped outside onto his doorstep one morning to find that half the population of his village had been wiped out by the great plague (in which half the population of Europe was destroyed). He wanted to call on God for help, but didn't know whether to address himself to Rome or to Avignon because two legitimate popes were at war. Not verbal war, *real* war! As he stepped from the threshold of his hovel, he was assaulted by a marauding band of mercenary soldiers who supported themselves in France by rape and pillage. This was the Middle Ages. Did the man know he was living in the Middle Ages?[4] Of course not. He just lived in Lyon. Did he know that the Renaissance was just around the corner? Of course not. No one did. Did he think the times were turbulent? *You bet.*

That brings me to another point about the confusion and depression in which we find ourselves engulfed. It breeds both dependency and magical thinking and creates an environment in which Technocrats thrive. Do the names Isaiah, Jesus, and Mohammed mean anything to you? Why is it that it has always taken *centuries* to produce a new prophet and this *decade* has produced a gaggle of "futurists" several thousands strong? Nostradamus would be so proud! This must be the first generation in all of human history to know nothing about the past and everything about the future. How can it be true that, whereas all previous generations had to struggle through without knowing the future, our generation does not? Isn't this just magical thinking? Aren't we just so insecure, so desperate, that we give a platform to anybody who claims to know where we're going?

What is this "science" of futurism if not the same one the Technocrat uses? Trend-line projection. You take a trend, say, world fossil-fuel reserves, and extrapolate it. Let's look at an example. Remember how the Club of Rome produced a dramatic report telling the world that the world's reserves of fossil fuel were running out? Pretty soon, they said, there wouldn't be a single drop left. They'd looked at a trend and projected it into the future. (The following graph illustrates my point.) Well, the whole world panicked. Governments scrambled to develop plans to deal with the emergency. In Canada, we had a National Energy Policy (NEP), which turned policy upside down, penalized our industry and our western provinces, and left permanent interregional scars. OPEC put the price up. High prices stimulated conservation and explo-

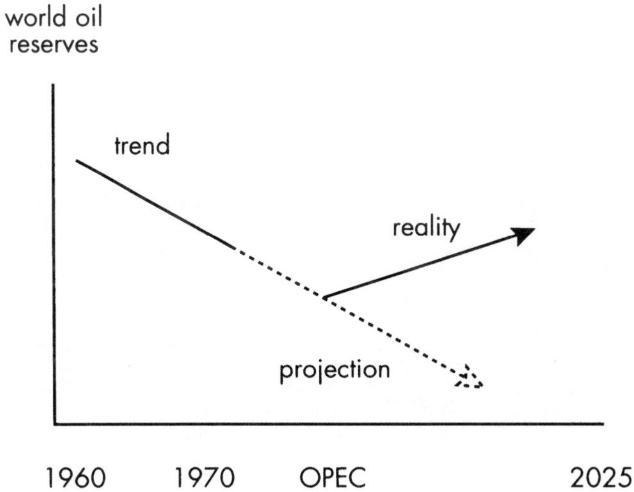

ration and substitution. Have you been in a gasoline lineup lately? Reality often surprises trend predictions. Sometimes the publication of trends succeeds in reversing them (like polls during elections, for example) and thus may serve a useful social function. To rely on them for knowing the future, however, is worse than hazardous. Benjamin Franklin said: "But in this world nothing can be said to be certain, except death and taxes."

When I get really down on myself, down on my capacity to run my own life in a satisfactory way, guess what I do? I read my horoscope! I figure, well, I can't run my life, so I might as well turn it over to some magical guidance from the planets. The more depressed I am, the faster I turn to it. But somewhere, deep in the back of my mind, I'm laughing at myself. I *know* this is a joke. I know I will emerge from the depression.

When we are depressed, we become dependent. We look for magical answers—drugs or alcohol, for example. In a social, collective context, we look to futurists (trend-predictors) and/or visionary leaders (trend-setters). We think the Technocrats have misled us—and they have—so we cry out for their opposites, the Artists. We want to flip from one form of dependency to another. This, I think, explains the tremendous upsurge in interest in leadership—and its perversion.

Unfortunately, we are, as usual, throwing a great many babies out with a lot of bathwater. I'll have a few things to say about that later, but for now, I want to explore the consequences of this play, the moral of this story, for organizations. To do that, I'm going to revisit some current notions about leadership, in Part 3.

Part 3

The Moral of the Story

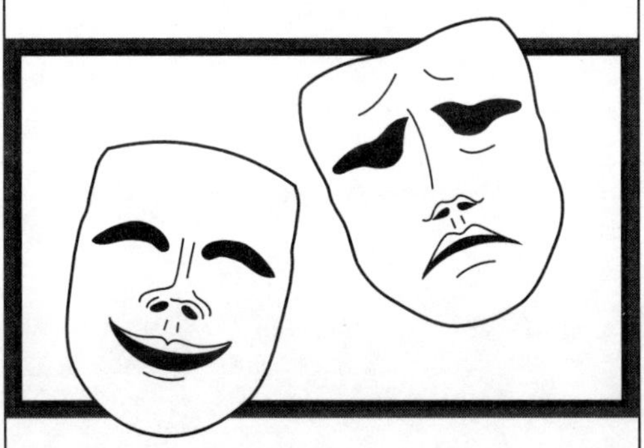

9

Leadership Revisited

> *To a person who knows his business as scientist, historian, philosopher, or any kind of inquirer, the refutation of a false theory constitutes a positive advance in his inquiry. It leaves him confronted, not by the same old question over again, but by a new question, more precise in its terms and therefore easier to answer. This new question is based on what he has learned from the theory he has refuted. If he has learned nothing, this proves either that he is too foolish (or too indolent) to learn, or that by an unfortunate error of judgment he has been spending time on a theory so idiotic that there is nothing to be learnt from it. Where the refuted theory, even though untrue as a whole, is not completely idiotic, and where the person who has refuted it is reasonably intelligent and reasonably painstaking, the upshot of his criticism can always be expressed in some form as this: "The theory is untenable as regards its general conclusions; but it has established certain points which must henceforth be taken into account."*
>
> R. G. Collingwood, *The Principles of Art*[1]

That's a very long quotation, but necessary, I think, because what I do not intend to do here is explain in great detail what a bunch of fools other analysts have been. Nor am I going to waste my time and yours on "idiotic" theories. Instead, I'm going to take a couple that make a lot of sense and see what I can add to them to make a little more. I'm going to take two of the best and most famous theories of leadership and go through the second one almost line by line.

Leadership According to Abraham Zaleznik

Almost 20 years ago, Harvard's Abraham Zaleznik created for us a distinction between *leaders* and *managers*.[2] To wake us up, he painted a stark contrast. Leaders, he argued, are visionaries; managers are planners. Leaders care about substance; managers, about form and process. Leaders inspire; managers motivate. Although his work became academically famous, to all intents and purposes it was ignored; his was a lone and courageous voice fighting a tidal wave of what he would later call the *managerial mystique*.[3] This mystique, with its pretensions to science, with its seemingly dispassionate "hollow men," with its obsession for process over content, for short-term profits, he claimed was responsible for the decline of American competitive advantage. America had stopped inventing, stopped dreaming, and was busy pushing paper around. It needed, desperately, to bring back leaders and get rid of all the managers. Zaleznik wrote:

> The managerial mystique is only tenuously tied to reality. As it evolved in practice, the mystique required managers to dedicate themselves to process, structures, roles and indirect forms of communication and to ignore ideas, people, emotions and direct talk. It deflected attention from the realities of business, while it reassured and rewarded those who believed in the mystique. . . . Essentially, business in America lost its comparative advantage by focusing on profits and stock prices instead of fostering innovation and long-term goals.

Sounds a lot like the Technocrat, no?

And Warren Bennis

Zaleznik, like many wise people, was ahead of his time. But time is catching up with him. Warren Bennis, easily America's foremost expert on leadership, confesses freely to a heavy reliance on Zaleznik's distinction between managers and leaders. In his 1989 book, *On Becoming a Leader,* he produces the comparative list below.

I claim that this list, however appealing, is misleading. Let me try to explain why.

1. *The manager administers; the leader innovates.* "Innovate: from Latin *innovare,* to renew, alter." My Artists certainly innovate on a grand scale. But Craftsmen innovate, renew—albeit on a smaller, more human scale—all the time. They change tools to suit the new circumstances. Technocrats innovate, too; they alter and claim to renew. Organizations go

The Manager	The Leader
Administers	Innovates
Is a copy	Is an original
Maintains	Develops
Focuses on systems and structure	Focuses on people
Relies on control	Inspires trust
Has a short-range view	Has a long-range perspective
Asks how and when	Asks what and why
Has an eye always on the bottom line	Has an eye on the horizon
Imitates	Originates
Accepts the status quo	Challenges the status-quo
Is the classic good soldier	Is his or her own person
Does things right	Does the right thing

through profound changes under their leadership. With our worship of change—radical change—we have tended to lend to the word *innovate* all kinds of positive connotations, but change is not always good and is far from always positive. The Technocrats triumph, in part, because they justify their actions in the name of change and innovation, and we worship at the feet of these idols.

2. *The manager is a copy, the leader, an original.* Message? Being an original is good (read: Artist), being a copy is bad (read: Technocrat). It is true, as we saw previously, that the Technocrats love to copy and that the Artists abhor it. But what about the Craftsmen? They are a bit of both. They work within traditions (copy) and innovate, experiment, risk at the margin (original). If we give vent to our animosity toward copying, we throw the Craftsmen out with the bathwater.
3. *The manager maintains; the leader develops.* I don't think so. The Craftsmen maintain *and* develop. The Artists develop while they let the Craftsmen maintain, and the Technocrats *undermine* maintenance and *pretend* to develop. When we condemn maintenance, in principle, and exalt development, we willy-nilly add fuel to the Technocrats' bonfire on which they roast the Artists and the Craftsmen.
4. *The manager focuses on systems and structure, the leader on people.* The Technocrats certainly focus on systems and structures. No complaint there. But the Artists? Do they focus on people? Not really. Basically they're very private, even solitary people, or at least oscillate between gregariousness and solitariness depending on their mood. They like people, but their focus is *outside* the organization, on dreams and deals. It's the Craftsmen who focus on people, their people. They look at the task and match it to the people. If they don't have someone to do the task, they train one of their trusted people to do it. They invest their time and the organization's money in someone. They prefer to invest in an existing employee who

already understands the business, because they believe that understanding the business is just as important as knowing how to accomplish the task. In fact, they believe that the two—knowing and understanding—are inseparable. Remember, we learned earlier from Polanyi that craftsmanship in general comes from this distinction between knowing and understanding. The craft of singing, for example, is "the science of acoustics *joined* to an experiential understanding of the body" (emphasis is mine). Try to imagine using just words—no singing—to teach someone who has no idea what singing is, how to sing: "Well, ah, you open your mouth and you push air out through your lungs . . ."

5. *The manager relies on control; the leader inspires trust.* Yes and no. The Technocrats certainly rely on control. They have to. They're scared. And the Artists certainly inspire excitement, but do they really inspire trust? Certainly they need it. They need trust desperately in order to have the time to complete their designs. But do they inspire trust in the whole of the organization? Sometimes. More often, however, they inspire trust among the Craftsmen in their immediate entourage; they're not in the habit of interacting much with people beyond their entourage. They're too busy dreaming and creating. It's the Craftsmen who inspire trust in the rest of the organization. It's the Craftsmen who engender loyalty ("He really went out on a limb to defend me and didn't even tell them that it was my fault!") and trust ("Tomorrow will be largely the same as today. I won't be fired on a whim or arbitrarily because the whiz kids are 'reengineering' ") and dedication ("He really believes in what we're doing; I'm proud to be a banker and to work with him") and self-confidence ("He let me cut my teeth on that project. He trusted me. I guess I'm competent") and humility ("Boy, I thought I had a brilliant idea, but he showed me why it wouldn't work. Oh, well, back to the drawing board").

6. *The manager has a short-range view, the leader, long-range.* Again, yes and no. The Technocrats certainly have a short-range view and are preoccupied by today's results. But with the help of futurism, those who once had only illusions of control now have the vocabulary of vision and the language of dreams. And they also have long-range plans that consist of a trend-line extrapolation into the future of their short-term successes. If you see their plans, you might think they have a long-range view, but they don't. The script reads something like, "Today, we have eliminated 30 percent of our inefficiencies (read: people) and we have achieved a return on investment of 10 percent; next year we will have eliminated 50 percent of our inefficiencies and will achieve a return on investment of 12 percent, and by 1995, 15 percent." Except that, as a Craftsman pointed out, "By 1995 we'll be dead." And they were.

   There's no doubt that the Artists have a long-range perspective; so long that they sometimes lose sight of today, or of how to get from today to tomorrow. That's where the Craftsman's medium-term vision comes in and is indispensable. Remember? The Craftsmen are the folks who build the bridges. James, Mike, and Cobb would never try to pretend that they could have built what they did without the able Craftsmen on whom they relied.

7. *The manager asks how and when, the leader, what and why.* Both the Craftsmen and the Technocrats ask how and when. Both the Craftsmen and the Artists ask what and why. The Craftsmen ask why when the Technocrats propose a new theoretical idea to fix something that isn't broken.

8. *The manager has an eye on the bottom line, the leader, on the horizon.* The Technocrats and the Craftsmen and the Artists all watch the bottom line. The Technocrats watch it for their glory, to show off, to show the world they're great admin-

istrators. The Artists watch it as a guide to the correctness of their long-term vision, to make sure they've picked the right things to do. The Craftsmen watch it carefully to *protect* the future. They know that if the bottom line is in trouble, the institution and its people will be in trouble. And yes, it's true that the Artists have their eye on the horizon, but it's the Craftsmen who see the hills and gullies that have to be crossed to get there.

9. *The manager imitates; the leader originates.* This is basically the same as the difference between a copy and an original in Point 2.

10. *The manager accepts the status quo; the leader challenges it.* Not entirely. The Artists certainly challenge the status quo in terms of their vision. But the Technocrats challenge the status quo with respect to the organization of work—constantly. There are new people and new organization charts all the time. One Craftsman told us, "They seemed to be fascinated with tinkering with the thing. You know. Systems and structures." The Craftsmen *respect* the status quo; they don't believe that everything we did in the past is necessarily bad or everything about the future is necessarily good. Their motto is "Change if necessary but not necessarily change." The more we give credence to the notion that change is always an unqualified good thing, the more we undermine the Craftsmen.

11. *The manager is the classic good soldier; the leader, his or her own person.* The Artist is certainly his or her own person, but if anything, the Craftsman is the classic good soldier: loyal, dedicated, respectful of hierarchy and chain of command. Think of General Colin Powell. The Technocrats are loyal to nobody but themselves. Not to the boss, not to the organization, not even to the industry. They'll jump ship if it will promote their own interests and they'll stab their own mother in the back if she gets in the way. The Craftsman is

also his or her own person; principled and honorable, not easily swayed or seduced. Think again of General Powell and why so many Americans wanted him to run for the Presidency.

12. *The manager does things right; the leader, the right thing.* The Craftsmen do both.

I hope I've made plain the nature of the problem. If reality consisted of only two types of people—the leader and the manager, or my Artist and my Technocrat—then who would have difficulty with choice? Not me. I'll take the leader any day of the week. But if I do, I must be prepared to make certain sacrifices, such as doing things right (Point 12), ignoring systems and structures (Point 4), control (Point 5), the medium-term (Point 6), and so on, sacrifices I am not prepared to make. In addition, I'll lose mentoring, a certain stability and continuity that is necessary for most mortals, organizational glue that comes from loyalty and dedication, and, in general, leadership that is characterized by realism and conviction, if not by vision. When we imagine a world polarized into these two camps, the manager and the leader, we lose the person in the middle. He or she becomes invisible. We long for Artists, but they are rare birds. In their stead, we get wolves in sheep's clothing—the Technocrats who, with their *beau discours* (beautiful talk), imitate Artists and Craftsmen. We make their con job easy by worshiping at the golden calves of Expertise and Change.

Why have I had to devote so much time to someone else's ideas about leadership? Well, because of that Collingwood quotation with which I began this chapter. You have to spend time on good theories, not stupid ones. From good theories, if you are not indolent, you can learn something. Second, everybody follows the best. Hervé Sérieyx, a French management consultant, has much the same thing as Bennis to say about leadership. So do all the minor, and not-so-minor, leadership consultants floating around hawking their wares: "You, too, can be a visionary leader. Just take five minutes and fill out this application form for my seminar." This leadership thing is the talk of the town, and I'm afraid the cure has become, if not worse than the disease, just as bad.

Obviously I sympathize with, and to a large extent share, the frustration we've all felt with Technocrats, or managers, if you prefer. (I no longer like this distinction between leaders and managers, because I prefer to say simply that there are good managers and bad managers.) Like everyone else, I'm sick to death of their pretensions and their sermonettes. I yearn for new inspirations. We want to get rid of the Technocrats and their inhuman, and inhumane, logic and bring back passion and creativity and imagination and experimentation. So, we have had to denounce them and show how wrongheaded they've been to imagine that life could be so easily controlled and planned and people turned into robots. But in order to undermine them, we are driven to an opposite extreme: We want passion and not logic, miracles and oracles and not plans and realism. But these are babies that I don't want to throw out with the bathwater.

Furthermore, as much as I love Artists—James and Mike and Cobb are delightful people—there is neither a need for Artists everywhere and at all times, nor enough of them to go around. In my organization, there were 4 at the top in the space of 60 years! Artists have always been a distinct minority, usually marginalized, most often denigrated. True visionaries are usually broke; that's what being a visionary means. No one believes you. Nostradamus wasn't called a visionary until people found some of his predictions in his grave. Jesus Christ was crucified. Galileo was put on trial. So was Socrates. As George Bernard Shaw said, "All great truths begin as blasphemies." If your favorite futurist is rich, take him with a grain of skeptical salt.

Finally, what are we going to do with all the rest of the people? The vast majority of whom are Craftsmen? Two stories to illustrate what I mean:

1. A young man, a 28-year-old MBA student, came into my office the other day. Here is the gist of our conversation.

> STUDENT: I was talking with professors X and Y and they sent me to you because you teach leadership. What I want to know is, what is it, this leadership thing?
>
> ME: Why?
>
> STUDENT: Well, I've been working for [a major telecommunications company] full-time and now part-time, while I do my MBA, for the last eight years, and every time I come up for evaluation, I don't do well in this leadership category. I ask them what they mean by it and no one can tell me what it is. I thought maybe you could tell me.

Well, what they mean is this: Are you inspiring, imaginative, visionary, creative, intuitive? I gave him a copy of Bennis's book. What I didn't tell him, what I didn't have the heart to tell him, is that hell will freeze over before he'll become a leader by this definition. He's intelligent and capable, eager to learn and to contribute, calm and collected, sincere and honest, but we don't value what he has to offer.

2. A young woman, 23, can't get a job. She has an advanced degree in management, is bright, attractive, and trilingual—French, English, and Spanish. She's good with numbers and with people. But she goes into an interview and some jerk inevitably asks her if she's a "leader." She's honest, so she says, "Well, sometimes. I mean . . . I mean, it depends. It depends on the circumstances." This is not what the interviewer wants to hear. He wants to hear, "You bet. I was the captain of this and that team. I got the best marks in school. All my friends look to me to tell them what to do. I'm very intuitive. I've been looking at your business and what you need is . . ." Confidence, right?

Ideas, right? Imagination, right? Modesty? Nope. Experience? Nope. Narcissism? Yep. Guess who replies this way in a job interview? A budding Technocrat. So, these interviewers who are looking for exciting young Artists are getting, for the most part, impressive, "brilliant" little Technocrats.

These kinds of interviews are going on all over the world at all levels of our organizations. Especially at the top. Let's listen in on an imaginary set of interviews with an Artist, a Craftsman, and a Technocrat.

INTERVIEWER: So what would you do if you get the top job?

ARTIST: I don't really know. We'll have to see. I have a few ideas, but they're sort of vague right now. I guess you'll just have to trust me. [*Remember what Cobb said?* "It's clear but fluid. Action brings precision. Very vague but becomes clear in the act of transformation. Creation is the storm."]

CRAFTSMAN: Well, we have to build on our strengths. I've been working here for the last twenty-five years and it's clear to me that whenever we've got off track with what we know best, we get in trouble. I think the widgets division has lost steam because the manager there doesn't know what a widget is. There have been so many changes that the sales staff is demoralized. There are no really big things that need to be done, just a lot of relatively minor things, experiments that will add up.

TECHNOCRAT: Now, I don't know anything about your business in particular, but all businesses have to decentralize, reengineer, empower people, emphasize quality, get into strategic alliances, become fleet-footed, supple, flexible, able to turn on a dime because the communications revolution is overtaking us, revolutionizing, globalizing everything almost overnight. Everybody is going to be a knowledge-worker operating out of his house or her car but linked into networks. The organization as we know it is going to disappear.

Who do you think gets the job? Who would you pick? The Artist is far too vague; she's a trend-setter not a trend-follower. She's a maker, a doer, not a talker—like Isadora Duncan, who cried, "If I could say it, I wouldn't have to dance it," or Robert Motherwell, who said, "All my life I've been working on the work—every canvas a sentence or paragraph of it." Since everything the Technocrat answers has become conventional wisdom, the Artist will be the last to employ it. She hates conventional wisdom.

The Craftsman is far too dull, and we're in a hurry.

The Technocrat mentions all the new management tools, but what he says is essentially vacuous. It would be like your contractor saying, "Well, I'm going to use screwdrivers and saws, and hammers and nails to build your house." You reply, "Yes, but what's it going to look like?" The Technocrat talks the talk. *"He multiplieth words without understanding."*[4] He seems imaginative. He seems farsighted. He reads the futurist magazines with their trend-line predictions. He sounds good. He hijacks the language of vision, but, of course, he won't decentralize or empower. He doesn't give a hoot about quality; he doesn't even know what you produce. Technocrats are the ones who read the management literature and go to the conferences, because they want to be on top of the vocabulary and they like conventional wisdom. In an earlier era, they were the ones who promoted the recipes of matrix management, management by objectives, and synergy; now, with the able assistance of institutionalized crystal-ball gazers and the pseudoscience of futurism, they masquerade as visionary leaders. They get the jobs. And that, ladies and gentlemen, makes me mad. And that's why I say the leadership cure is getting worse than the technocratic disease.

Neither Abraham Zaleznik nor Warren Bennis intended this to happen. Both of them want real Artists, real leaders, not imitators. But you cannot control what others will do with your work, nor can you control the climate of opinion in which it will be received. Zaleznik and Bennis are not responsible for futurism's hold on us. And while Bennis seems ambivalent about it (note "becoming" in the title of his book, *On Becoming a Leader*), I feel safe in saying that Zaleznik certainly doesn't believe that anyone can *learn* to be a visionary leader, lateral thinking notwithstanding. But because we have polarized the world into two camps, the bad guys (managers) and the good guys (leaders), and because we have popularized the idea that anybody can become a good guy with enough training or desire, we have unwittingly lent the Technocrat more ammunition to drive out the Craftsman and the genuine Artist.

Craftsmen represent the vast majority of people, young or old. So I ask again, what are we going to do with all the people if we now value only Artists, or what Sérieyx calls *l'acteur/auteur* (actor/author), at all levels of the organization?

> WANTED: Volatile, emotional, unpredictable, imaginative self-starter. All others need not apply.

Oh, you say you don't want *emotional, volatile,* and *unpredictable,* just *imaginative?* Sorry, they only come in a package: character. I can offer you a dedicated, loyal, honest, realistic, knowledgeable package, but the imagination bit will be rather limited. Do you really need a lot of imagination, or will a little do?

Collingwood told us at the beginning of this chapter that the upshot of examining good theories can "always be expressed in some form as this: 'The Theory is untenable as regards its general conclusions; but it has established certain points which must henceforth be taken into account.' " In my view, the theories we have seen here are "untenable as regards their general conclusions." Let's turn now to a discussion of "certain points which must henceforth be taken into account" when we're thinking about what organizations might really need.

What we've learned from Zaleznik and Bennis, and what must henceforth be taken into account, is what we *don't* want, what leadership *isn't.* We don't want rigidity. We don't want an exclusive preoccupation with today at the expense of tomorrow. We don't want people to be treated like robots. We don't want to suffocate in systems and rules.

We want vibrant, interesting organizations. We want organizations with enough flexibility to adapt to an as yet unknown and, I claim, unknowable future. But those organizations also have to be stable enough to provide us ordinary people with a sense of continuity, without which we feel lost. We want experimentation. But we also want the tried and true. We want long-range perspective, but not at the expense of losing contact with the present and the immediate future. We need both to respect and to challenge the status quo by going back and by going forward. We need to do the right thing and to do things right. We want and need true visionaries who discover new paths, but we can't expect them to make all our struggles disappear. They are not our saviors. We need to take charge of our own fates, to accept responsibility. And, finally, we need very badly to be able to recognize charlatans, con artists, and false prophets, and we need to stop bowing down before the golden calf of Expertise, which pretends to combine Competence and Virtue.[5]

We need to *recognize* (*re-cognize:* to resume possession of, fifteenth century; to know again, sixteenth century) character. We need to be able to see the Artist, the Craftsman, and the Technocrat, not as rigid categories, abstract concepts, or neat little boxes, but as *central tendencies* of individual human beings. Although there are some examples of pure types—we have already met Cameron, who is about as pure a Technocrat as you can get, and we will meet some more in the following pages—people are predominantly one style with some minor elements of another; sort of like your astrological sign, "Virgo with a rising Cancer." Following is the list of central features of the archetypes.

In my research, I found nine basic configurations of these lists (plus some adjectives which fell between two). Let's follow the map below.

Artist	Craftsman	Technocrat
Unpredictable	Well-balanced	Cerebral
Funny	Helpful	Difficult
Imaginative	Honest	Uncompromising
Daring	Sensible	Stiff
Intuitive	Responsible	Intense
Exciting	Trustworthy	Detail-oriented
Emotional	Realistic	Determined
Visionary	Steady	Fastidious
Entrepreneurial	Reasonable	Hardheaded
Inspiring	Predictable	No-nonsense

Technocrats

Pure Technocrats

Cameron is the best example. The qualifications you're looking for are *cerebral, uncompromising, stiff, hard-headed,* and *detail-oriented.* Cameron is cold as ice; as I noted earlier, he scored zero on *people-oriented, amiable, warm,* and *generous.* He's calmly superior. Totally humorless. Couldn't laugh at himself if his life depended on it. He will brook no opposition. In fact, he acts as though there couldn't possibly be any analysis which differs from

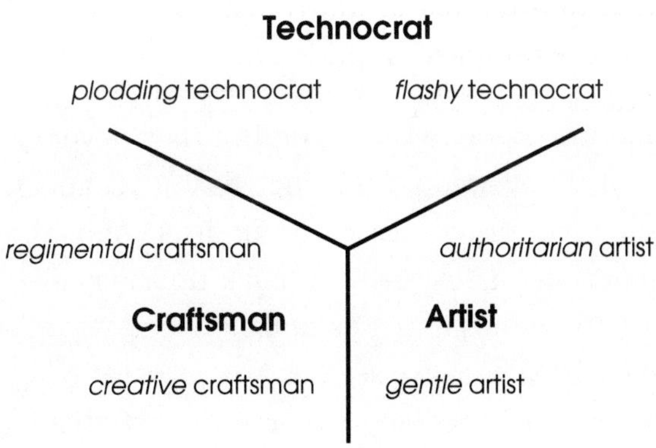

his own. That's why I say *calmly* superior. All the Technocrats feel superior.

Plodding Technocrats

Take the plodding Technocrat, for example—somebody like Brien or Judd. This nuance also came to me from someone in an audience. A woman came to me and said, "Can you be a Technocrat without being brilliant? My boss conforms to everything else you describe except that." Sure. Not all Technocrats are brilliant, they just all think they are! Brien and Judd both thought they were the cat's meow. The plodding Technocrat is rigid, dogmatic, uncompromising, and hard-headed—central technocratic tendencies—but not all that bright. It's the quick-witted ones who are the most dangerous and the hardest to recognize—the flashy Technocrats.

Flashy Technocrats

People like Ross and David. They're hard to describe. As usual, it was a Craftsman who gave me a hint and an example. During an interview with one of the Craftsmen in the study, he was filling out an adjective list on one of the Technocrats. He had checked off *bold* and *conservative.* Since *bold* had always been associated with the artist and *conservative* and *bold* are contradictory to me, I asked him to explain. He said, "It's simple. If you're bold about an old idea, you're still conservative. If you're bold about a new idea, you're an entrepreneur." The technocrats were "bold" in the sense of forceful and determined. They followed the cutting costs part of the reengineering vogue to the letter. But they never came up with a new idea. They never can. But, since flashy Technocrats are also *brilliant,* they can argue circles around us. And, they'll usually have 16 studies and 4 "experts" to back up their point of view. If you can have 16 studies and 4 established experts on a subject, it has to be an old idea because studies are necessarily based on past data and experts only become experts after studying an existing phenomenon. These fleet-footed Technocrats are often confused with Artists of the authoritarian variety.

Artists

Pure Artists

Cobb is the best example. Here the qualifiers are *unpredictable, imaginative, daring, intuitive,* and *exciting.* Cobb is the man I earlier compared to a thoroughbred locked in the starting gate. He's fidgety, full of nervous tension and energy. He's always running. He produces an idea a minute and that's why one of the Technocrats called him "*Star Treky.*" New ideas make Technocrats very annoyed—in part, I think, because they're jealous.

Authoritarian Artists

The authoritarian Artists are often hard to get along with. They're not great listeners, they're brilliant and very determined and often highly analytical. But, while the flashy Technocrat pedals old ideas and says "It's my way or the highway," the authoritarian Artist says "If you don't like my dream, build your own." People buy into their dream, if not into their personality. Lots of subordinates are motivated by the idea itself and are prepared to put up with the personality. If you want personality, get yourself a gentle Artist.

Gentle Artists

Gentle Artists are soft-spoken and usually self-effacing. They're generous and warm, easy-going and people-oriented. James was like that. An Artist, with a minor in Craft. Big dreams, big projects, but no big talk (not like a Technocrat, who is all big talk) and no big ego. He lived with lots of doubt, did not take himself too seriously, never looked back on failures, and never blamed anybody but himself for them. That's why he's heartsick about his nomination of Cameron. He doesn't blame Cameron. He blames himself and his own bad judgment. Sometimes the gentle Artist can be confused with the creative Craftsman.

Craftsmen

Pure Craftsmen

People like Robert and Jeb. Robert was the one who wanted to spend the greater part of his work year going around to all the branches. Robert is steady, reasonable, realistic, and trustworthy. He's calmly intelligent and reflective. He's tolerant of human error, patient, and usually gives people the benefit of the doubt—sometimes when he shouldn't. He's knowledgeable, knows the business because he's been in it all his adult life. The only time he becomes impatient is when someone tries to tell him how the business *should* work in theory when he knows perfectly well that it doesn't in fact. No one would ever call him exciting, but normal people love working with him because he genuinely values their contribution.

Regimental Craftsmen

The regimental Craftsmen are very methodical and analytical, serious, quite conservative and controlled, but they *share* with the pure Craftsmen their open-mindedness. It just takes more to persuade them to change their mind. Rodney, for example, described an incident that happened to him in the early 70s. It was at the time that encounter groups were all the rage in business circles. His boss had planned a day-long session, and he was supposed to attend. He went, all right, but he left after 30 minutes. "Sitting around talking about your so-called inner feelings! I couldn't take it. It was just a lot of hogwash to me, but I know some people take it seriously." You see, Rodney is very reserved, private. But—and this distinguishes him from Technocrats—he does not *impose* his version of reality on others. He didn't prevent his staff from attending. Recently, after making a speech, a 76-year-old man came up to me and said, "During my career, I was pretty methodical and detail-oriented. Does that mean I'm a Technocrat?" I replied, "If you ask that question, you aren't one." You see, Technocrats, no matter what their stripe, never think there's anything wrong with them.

Creative Craftsmen

Creative Craftsmen share a self-effacing demeanor with the gentle Artists. They're amiable and generous but—also like pure Craftsmen—realistic, sensible, and team players (whereas Artists of whatever stripe are essentially loners). What perhaps best distinguish the creative Craftsmen from the gentle Artists are two things. One, while Artists come up with a new idea about every five minutes, creative Craftsmen come up with one about every five years. It's an idea in *continuity* with the past and not a radical rupture. Two, they build the germ of an idea slowly but surely by getting everyone to buy into it, refine it, amplify it, and make it happen.

While we would like to combine all three—the analytical brilliance of the Technocrat, the imagination of the Artist, and the wisdom of the Craftsman—we can't. Human beings aren't gods. Only gods combine all three, and only technocrats think they're gods. So, it turns out, Herbert Simon *was* wrong. There is no amount of training, no management course, no management book that can turn a hard-nosed arrogant Technocrat into a wise Craftsman, or an authoritarian Artist into a listener; nor is there an "expert system" that can turn a gentle Artist into a Technocrat. Management always has been and always will be about the judicious selection of the right people for the right job and not primarily about training.

As James now knows all too well, if you put the wrong person in the wrong slot, the whole project can collapse in the wink of an eye. As poet Robbie Burns told us, *the best laid schemes of mice and men . . .*

These types are crystal clear for me. I've known these men for a decade. But you don't know them, so now I'm going to turn to some examples you may know better. I know many of us feel that the whole of life has been taken over by the Technocrats, but that's not true either. How many are there, and what do they look like? The first step in the resolution of any problem is its recognition. Here come the only recipes you will find in this book: How to recognize Artists, Craftsmen, and Technocrats.

10

Where There's Smoke, There's Apt to Be Fire

At last the secret is out, as it always must come out in
the end,
The delicious story is ripe to tell to the intimate friend;
Over the tea-cups and in the square the tongue has its
desire;
Still waters run deep, my dear, there's never smoke
without fire.

W. H. Auden, *Twelve Songs VIII*

As I said, it feels like technocrats have taken over the whole of life, but have they really? On the theory that where there's smoke there's apt to be fire, let's take a look at some recent history and see if we can see signs of the different types at work in other organizations. In what follows, I rely on many secondary sources, in particular on an excellent book by James Collins and Jerry Porras, *Built to Last.*[1]

On the Corporate Landscape

Technocrats?

Collins and Porras describe in some detail the different evolution of two competitors, Wal-Mart and Ames. The Gilmans, founding family of Ames, had a head start on Sam Walton and achieved considerable early success. But, whereas Walton built a corporation dedicated to innovation and employee creativity, and groomed a successor who believed in the "Wal-Mart Way," the Gilmans apparently made a fatal succession error, much as James did in my story.

> Ames leaders dictated all changes from above and detailed in a book the precise steps a store manager should take, leaving no room for initiative . . . Ames CEO of the same era said, "The real answer and the only issue is market share."

Doesn't this sound like technocratic leadership to you? Precise, detailed rule books which presume that people are robots, and the single-minded pursuit of the conventional wisdom of the day—"the only issue is market share." Building market share was seen by the *experts* as the panacea of the day, and if you didn't buy it, you must be a fool or an imbecile.

Another story from another sector—this time, Philip Morris versus R. J. Reynolds. Collins and Porras tell us:

> . . . during the 1970's and 1980's, while Philip Morris relentlessly invested in its goal . . . R. J. Reynolds executives used the company primarily as a platform for their own self-aggrandizement and enrichment. They bought a fleet of corporate jets (referred to as the RJR Air Force), built expensive airport hangars (referred to as the "Taj Mahal of corporate hangars"), constructed elaborate and unneeded corporate offices (called the "Glass Menagerie"), decorated with expensive antique furniture and exquisite art work . . . When asked about the wisdom of these expenses, CEO F. Ross Johnson responded simply: "A few million dollars are lost in the sands of time."

Now I wouldn't dare claim that Ross Johnson is a Technocrat—I don't know the man and I haven't studied him—but I can say this: There are striking similarities between this description and my Technocrats. While James was content with a modest head office building in his home town, his successors built a corporate "Taj Mahal" in the financial capital, decorated with expensive artwork, and they, too, had their fleet of corporate jets. Flitting around the world in these jets made them feel like big shots, visible proof that they were very important people indeed. Furthermore, whenever anyone dared raise a question about these expenses, they, too, were dismissed. Where there's smoke . . .

And how about this one? Howard Johnson versus Marriott:

> While Marriott continued to invest and build for the future even during recessions, Howard Johnson became overly focused on cost-control, efficiency, and short-term financial objectives. While Marriott pushed itself to continually improve the quality and value of its service, Howard Johnson became "overpriced and understaffed purveyors of pallid food, hamstrung by outdated ideas." "Profits were artificially high. Its reinvestment had been neglected. Pennies had been pinched on staffing, menus and renewal. It was milking the business by not reinvesting." At one point, Johnson Jr. moved to elegant quarters in New York's Rockefeller Center . . . and spent the bulk of his time socializing in elite society. Summed up a competitor: "Every time I saw Howard Johnson he was always telling me how he was going to cut costs. I don't think he spent enough time at his restaurants. If he'd eaten at his own restaurants more instead of lunching at 21 [a fashionable New York restaurant], he might have learned something." In contrast, Marriott Jr. lived a relatively modest lifestyle guided by what he calls the "Mormon work ethic" that drove him to personally visit up to two hundred Marriott facilities per year—and to expect similar travel schedules from other top managers.

What's technocratic about that, you ask? Well, let's see. All organizations need to concentrate on "cost-control, efficiency, and short-term financial results." That's true. But my Technocrats concentrate only on these things to the exclusion of everything else. To the exclusion of marketing, to the exclusion of R&D, to the exclusion of employee training, to the exclusion, in short, of the future. "Profits were artificially high, reinvestment had been neglected" in my company, too. As one artist said, "They've picked all the fruit from the tree to artificially inflate earnings. Now what will they do?" They squeezed the lemon to impress "the Street," to show what great managers they were—and that works for a while, because all those costs squeezed out of the system go straight to the quarterly bottom line. But, now what will they do? I'll tell you what. They'll bail out before the long-term damage is too visible and go on to greater glory somewhere else on the strength of their so-called great management skills.

So, while Johnson is eating at 21, Marriott is schlepping around the country visiting his restaurants. Do you remember when one of my short-lived Craftsman CEOs told his boss, Cameron, that he intended to visit all his branches in the coming year? Do you remember what Cameron told him? "Stay home. You have more important things to do." That is, do studies, cut costs, drive profits to the quarterly bottom line. That's what.

Last, but by no means least, wanting to be seen in "elite society" is a special feature of Technocrats. They love to preen in public. They're narcissistic. Of course, maybe Johnson Jr. isn't personally a Technocrat. Maybe he just left his management team in the hands of other Technocrats. Maybe they were all great guys. Maybe they relied on bad consultants. Maybe . . .

Artists?

Unless their business is words, like a poet, you recognize Artists by their deeds, by their track record, not by their words. If they use words, the words will surprise you; you won't have heard them before in quite the same way. They're unpredictable, remember? Just when you think you've got some subject truly mastered, they'll come along and push your head around, wake you up, make you question your certitudes. As soon as there is a bandwagon for something, they'll get off and create a different one. They never brag; they doubt their own capacities even if they don't voice those doubts to you. What we're looking for is a *continuous pattern* of creativity, of concrete accomplishments, and not just ideas (flashy Technocrats talk a good fight, but if you look at their track record they've never accomplished anything risky, unconventional, or creative). Let's see.

How about Ted Turner of CNN? What's he done (as opposed to said)? A *continuous pattern* of risk-taking and bold gambles all of his adult life. Here's how Landrum describes him in *Profiles of Genius*:

> Ted Turner epitomizes the consummate risk-taker in business. In 1970 he risked his billboard advertising business to acquire a failing UHF TV station, WTBS. He then risked the station to acquire the Atlanta Braves and the Atlanta Hawks. Next he risked the total $100 million organization to create CNN. Then he risked everything in an attempt to acquire CBS and nearly lost it all in his successful acquisition of MGM.[2]

What's he like personally? Landrum continues:

> Dan Schorr of CNN said of Turner, "He was a bundle of energy. He never stopped moving, almost with an animal quality, like a tiger, never not in motion . . . Ted was also described as having the attention span of a gnat. CNN personnel have said, 'Holding a conversation with him is like trying to talk to a radio.' "

Doesn't this remind you of my description of Cobb as a nervous thoroughbred before the big race? A bundle of nervous tension and agitation.

And there's more. Other similarities. Apparently, Ted Turner was a hard drinker and hard player, just like Cobb. Highly intuitive; one of his subordinates said of him, "If Ted predicted the sun will come up in the west tomorrow morning, you'd laugh and say he was full of it. But you'd still set the alarm. You wouldn't want to miss the miracle." Cobb's vision was also 20 years out; people never understood where he came up with this vision and neither did he. He told us that the vision comes from "astrology, quirks, dreams, love affairs, science fiction, perception of society, some madness probably . . ." What's the madness?

It's called manic-depression. Ted Turner is on lithium, the preferred medication to treat it. Like Cobb and Mike and James, who are cyclothymic (a mild form of manic-depression), Turner has apparently had to live with periods of deep depression followed by manic energy and creativity and acquisition binges—binges of all sorts. His father committed suicide, and we know that depression and manic-depression run in families.[3] There's a genetic base. Severe manic-depressives are incapacitated and often commit suicide. Fortunately, Ted has escaped that fate.

And how about another one? Walt Disney. What did he do (not say he would do)? What was he like? We all know that Walt Disney's whole life was characterized by a *continuous pattern* of creative productions. What many don't know is that he had a nervous breakdown in 1931; or that he was obsessively, morbidly, preoccupied by death; or that his father was notoriously unstable, a depressive, and often violent. Walt himself was solitary; apart from his family, he seemed to have had no close friends. Highly imaginative, visionary, entrepreneurially determined, even obsessed, he was a perpetual motion machine. His daughter describes him as constantly agitated:

> A conversation with Father is an experience. At times mere words won't express what he's trying to say, and as he talks, his hands move incessantly. His gestures and his facial expressions are as much a part of the conversation as his speech. . . . He looks at you . . . but he's so lost in his enthusiasm, I'm not sure he really sees the one to whom he's talking . . . I've seen him carry on a one-sided conversation for a long time and come away convinced that the one who has been listening is intelligent, sensible and highly articulate—although the listener hasn't said a word.[4]

And Walt was totally unpredictable. His daughter continues:

> It's that insatiable, omnivorous curiosity of his that keeps him going . . . It will keep him changing Disneyland and everything else he can, until there's no space left for anything new: then he'll rip out something old so that there *will* be space. No one, not even those closest to him, can anticipate what he'll do next, but there'll always be something, and no matter how impossible it sounds, Father will do it (p. 274).

Solitary, unpredictable, imaginative, visionary, entrepreneurial, periodically seriously depressed and then exuberant, family history of depression and emotional instability. How much smoke do you need?

And what's Bill Gates of Microsoft all about? A *continuous pattern* of creative activity over his (relatively short) lifetime. He's described as brilliant and fidgety and driven: What's he running from? Depression maybe? He admits living with a perpetual fear of failure. He sure feels like an authoritarian Artist to me. And how about Stephen Jobs of Apple and Next? Difficult, sometimes autocratic and erratic, spectacular successes (and some failures, but I wouldn't consider him down for the count). He spent most of his youth looking for gurus and answers to life's emptiness. Who feels that life is empty? Depressives. How do depressives combat it? With dreams. Big Dreams. Big Projects. Big Hairy Audacious Goals.[5] Now, maybe none of these guys are artists. As I say, I don't know them personally, and I haven't studied them, but what do you think? I see a lot of smoke and where there's smoke . . .

Craftsmen?

How do you recognize them? Dead easy. Intelligence without shine and polish and flashiness. Ethical and principled. Frank. Wise. Well balanced. Experienced. Knowledgeable. Reasonable. Good judgment; they can see the Technocrats a mile away but they respect dreamers: "The analytic boys are a dime a dozen, but you can't buy dreams." Look around you. They're everywhere, but you probably don't think of them as leadership material because they're not flashy. You may even think they're boring. Listen to a conversation I had recently:

HER: I've been here three months and I'm in a state of shock!

ME (fearing the worst): Why?

HER: Well, before I came, my future boss made all kinds of promises to me about how I'd be able to operate. But, where I'd come from I had heard all these before so I was very skeptical. But, he's doing it! Everything he said was true! He's living up to his promises! What a shock! He said I would have a lot of autonomy and I do! He said he'd get me money to try new things, and he does! Gee, it almost makes you a believer again. I didn't think there were any like him any more!

ME: What's he like personally?

HER: He's very calm. When we get overwrought and worried, he calms us down. He's very honest. Straightforward. He doesn't make promises he can't keep. Today, for example. He's not here. He promised his wife to do something with her today and in spite of all the pressures, he's doing it! He leads a very well-balanced life. You know, sane. And the most amazing thing, I'm almost embarrassed to say it, you'd almost say he has scruples! I didn't think anybody in power had scruples anymore!

Now, *calm* is not the first adjective that springs to mind for a so-called visionary leader, an Artist. Nor is *well-balanced.* And, while Artists usually have scruples, it's the Craftsman's badge of honor. Do you think this woman will work hard for that man? You can count on it. Do you have to be "visionary" to motivate people, to inspire them? Nope. Some people are inspired by honesty and conviction and sincerity. Does that mean that Craftsmen are not *leaders,* they're *managers?* I don't think so. Nor does this woman. And that's how this manager/leader thing gets us into trouble.

There are different ways of leading, and we've come to imagine that if you're not charismatic and visionary you're not leading at all! Porras and Collins have written a whole book that talks about Craftsmen at work, but they don't know it. They're so annoyed by the idea of visionary leadership that they talk about "visionary" companies and wash out the very notion of leadership. The companies they describe—like 3M and Merck and GE—are not "visionary," they're just good. And that's not accidental. And they have had leaders. Everything they have done has been done by somebody, some person. Companies do not run themselves. Their leadership seems, in the main, to have been Craftsmen. Let's take a close look at some of the things Porras and Collins are saying:

Charisma Not Required

> . . . great leadership, be it charismatic or otherwise, cannot explain the superior trajectories of the visionary companies . . .

Oh yes it can, but not if you define leadership by *charismatic.* Let's look at one of their best examples, 3M. Porras and Collins write:

> Consider William McKnight. Do you know who he is? Does he stand out in your mind as one of the great business leaders of the twentieth century? Can you describe his leadership style? Have you read his biography? If you're like most people, you know little or nothing about William McKnight . . . Yet the company McKnight guided for fifty-two years [from 1914 to 1966] earned fame and admiration with businesspeople around the world; it carries the revered name Minnesota, Mining, and Manufacturing Company (or 3M for short). 3M is famous; McKnight is not. (p. 32)

Now this is meant to prove that because McKnight himself was not a visionary, visionary leaders are not necessary. Just build a great company and the rest will look after itself. But, who builds great companies? Do they build themselves? What was McKnight like?

> . . . a soft-spoken, gentle man . . . a good listener, humble, modest, slightly stooped, unobtrusive, quiet, thoughtful and serious.

Sure sounds like a Craftsman to me.

The Company Itself Is the Ultimate Creation

> Be prepared to kill, revise, or evolve [*sic*] an idea . . . but never give up on the company . . . if you see the ultimate creation as the company, not the execution of a specific idea or capitalizing on a timely market opportunity, then you can persist beyond any specific idea—good or bad—and move toward becoming an enduring great institution.

This is meant to show that if you focus on the company itself, as an institution with a life of its own and an importance beyond the personalities of its current leaders and current products, you'll come out a winner in the long run. I, of course, have no quarrel with this in principle, but, again, I have to ask *who* treats the company like an institution? What kind of people put the institution above and beyond themselves? Rodney, one of my craftsmen, told us earlier, ". . . the institution was important. People like to feel that they're part of something important, have a sense of belonging, *pride.*" If you're a Craftsman, you don't have to be told to treat the institution with kid gloves; you do it instinctively. You're modest and you see yourself endowed with a responsibility to ensure the long-run survival and prosperity of the institution itself. You have a trust to keep. The organization is not the instrument for your self-aggrandizement. That's who you are. It's your character. Now, you can tell Technocrats to do this and they will probably tell you they will. But they can't and they won't, so save your breath. They are in the game for their greater glory; not yours and not the company's. And furthermore, while many Artists are heavily identified with their corporate creations and want the institution to outlive them as a monument, they're more interested in the dream for which the company is but the instrument. They regularly bet the institution as a whole, risk its demise, in order to pursue the dream. We saw this very clearly with Ted Turner. Craftsmen do not risk the demise of the institution to pursue a dream.

Hire from Within

> ... across seventeen hundred years of combined history in the visionary companies, we found only four individual cases of an outsider coming directly into the role of chief executive.

Answer? If you want a good company, hire from within. Question? Who hires from within, who believes most in continuity? *Craftsmen.* As all the experienced people were being fired in my organization, to be replaced by whiz kids from outside, the Craftsmen knew that it would spell disaster. They knew that continuity was important. And, the Technocrats certainly had a habit of hiring from outside, but it's *who* they hired that mattered most—not *where* they hired. They hired their clones. They then inaugurated elaborate succession-planning schemes, paying an outside consultant hundreds of thousands of dollars to analyze managerial capacities up and down the organization. When promotions came up, they went to other Technocrats. You see, it's not the principle of hiring from within that is critical. It helps, to be sure, to ensure a sense of permanence. It persuades people that they are valuable to the institution. It sends healthy signals right to the bottom. But the presence or absence of succession planning per se tells us nothing. What principles and policies guide hiring and promotions? Do you have to be dedicated or a brilliant analyst? Do you have to be a team player or a loner? Do you talk more than you listen? And so on, and so on, and so on. If these great companies picked good leaders consistently, it's because they had good judgment consistently. Who has consistently good judgment? Craftsmen.

Do Artists necessarily have good judgment? If you can catch them at the crossover point between depression and mania they have excellent judgment, but it's really not their forte. *Built to Last* is riddled with examples of Artists making very bad succession decisions (or leaving it to chance). In my case, James chose Cameron. Either because they're insecure about their own talents and therefore choose their opposites (did Jobs choose Sculley?), or because they're too authoritarian to keep anybody good around them, or because they're so afraid of dying that they tell themselves they'll live forever and won't need a successor, or because they're too busy dreaming to think about succession, they often get it wrong and the organization pays the ultimate price.

So, as I said, Collins and Porras have written an excellent book based on some solid research and it does what good academic books ought to do: It describes for us the "best practices" of excellent companies. If I were a CEO, I'd want to read it to get some ideas. But—and there is a but—they tell us, essentially, get the systems right, get the core values right, get yourself some BIG HAIRY AUDACIOUS GOALS (BHAGS), hire from within, and everything will look after itself. But wait a minute. James had the systems right, the core values right, the BHAGS right, and he hired Cameron from within, but *he hired the wrong guy* and everything went down the tube. In the flash of an eye. As Ward Melville of Melville Corporation said, "I was shocked to see how fast the numbers can deteriorate when the men are wrong."[6] James promoted a pure Technocrat—cold, distant, stiff, arrogant, uncompromising, and cerebral. Technocrats don't have values, core or otherwise; they fulfill Oscar Wilde's definition of a cynic: someone who knows the price of everything, and the value of nothing. And Technocrats don't like BHAGS. They like SMARTS: sophisticated, measurable, annual, rigid targets—like ROI.

Let's have a look now at what organizations really need: teams. People with different talents and skills working hard together.

11

About Teams

Yet it was observed that these mechanistic explanations always left Hamlet out of their play; that, in brief, like all merely mechanistic explanations, they left out the vital spark or soul or essence—the essential mark or quality or character that made the phenomenon what it was. . . . Whatever well-formulated and even demonstrable "truth" these scientific explanations had, they had obviously no "reality" nor any profound relation to the living experience of any man. . . . They told us many interesting superficial things. They never uttered the secret.

Ludwig Lewisohn[1]

Although we hear a lot of talk these days about something called the *Learning Organization,* organizations don't learn and they don't see and they don't do anything. *Organization* is a concept, a shorthand; it is the name we give to a certain collectivity of people. It is a useful way to be able to talk about these collectivities without having to say every time, "Such and such a group of people organized by some principle to attain some objective or other." It is a useful *abstraction* for some purposes, but not for all. Abstractions, concepts, neither learn nor see; people do. North American business has been accused, for example, of taking a short-term view at the expense of the long-term. Is this, strictly speaking, true? I don't think so. North American business is an abstraction; it doesn't exist. It doesn't take long or short views. North American businesspeople are *real.* If North American business has taken a short-term view it is because North American managers have taken a short-term view. And by now it's clear, I hope, that not *all* North Americans take a short-term view. The Artists usually don't. The Craftsmen usually don't. The Technocrats do, and we have given them free rein. That's not their fault. It's ours. So let's own up to our part in their triumph. We should have known better, but we didn't. No point crying over spilt milk. What we can do about it now is what matters.

Since organizations are composed of people, it seems to me not farfetched to imagine that what may apply to people in general may apply equally to organizations as a whole. Since every organization faces its own unique challenge, there are precious few things they all need, very few universals. Just basic things—a team of people working together to promote the three Es: Efficacy, Expertise, and Efficiency.

Efficacy: On Doing the Right Thing

Efficacy is, to paraphrase Peter Drucker, the dean of American business commentators, doing the right thing. Efficacy is making the right choices about goals. How do organizations go about setting goals or strategies? There have been, essentially, two dominant schools of thought about creating organizational mission: the rational/analytic and the experiential. The *rational/analytic* view insists that you determine an appropriate strategy by collecting a lot of information about your markets, your competitors, and your suppliers, and you, in effect, *calculate* your strategy. You develop a *plan*. This notion of strategy as plan is, of course, an old and creditable one. Plans are good. However, in the realm of human activity, it's tough to plan. The planning mode is based on the assumption that the future will mimic the past. This is so because analysis always comes late to the party; it is based on collecting historical "facts," what used to be, and extrapolating (projecting trend-lines) those facts into the future. One problem with this approach is that the facts need interpretation and only the Technocrat thinks that the facts speak for themselves. A second problem is that change is unfathomable, and trend-line extrapolations are seldom accurate. (Haven't we been given trend-line projections that purport to prove again and again, year after year, that our governments' deficits are going to disappear by 1992, 1993, 1995, 1997, 2000?) At one time they told us that you couldn't have inflation and unemployment at the same time (the Phillips curve), but we learned the hard way that that particular trend didn't sustain itself.

What if everything defies the trends? As Cobb told us, things always change—"government, death, technology." In 1986, Cobb tried to explain to me a complicated set of financial maneuvers that would allow his company to swallow an organization twice its size. I didn't understand a word he said except the name of the target company. Five years later, I sat in a meeting and listened to his successor talk about the results of a planning exercise—for which he had paid a consultant $400,000. The analysis came up with the same target company as Cobb had! Trouble is, it was *five years too late.* Everything had changed. The trends were not borne out. The group had no money, and the window of opportunity had closed. So, it's not that analysis is bad; it's that it's slow. It does not lend itself very well to prediction, to determining long-range goals.

Frustrated with what was seen as this lack of realism in the rational/analytic, blueprint model of strategy-making, management scholars like Henry Mintzberg and James Brian Quinn turned to a more *experiential* model. Quinn told us that in the real world, strategy does not "spring full-blown like Minerva from the brow of Jupiter," because it "involves forces of such great number, strength and combinatory powers that one cannot predict events in any probabilistic sense. Hence, logic dictates that one proceed flexibly and experimentally."[2] He called what we needed *logical incrementalism.* Mintzberg would add his eloquent voice to this in his award-winning *Harvard Business Review* article, "Crafting Strategy." And Tom Peters would reach fame and fortune by celebrating experimenting, acting, trying things. On the other side of the Atlantic, Hervé Sérieyx would argue that strategy is out and tactics are in.[3] Quinn hinted at the assumptions behind this glorification of experimentation. We cannot predict the future. We are like the blind. We cannot see through the dense fog of turbulence and complexity; we must use a blind person's cane to tap on objects around us.

Well, all right, I can't see through the fog, but some people can. Artists can. To make en*light*ened choices (note my emphasis on *light*), we need to be able to see. We need vision. Now, because we have seemed as a society to have lost our way, we cry out for vision and for visionary leaders as though such people could provide miraculous answers for us. This is, I think, the reason for the dramatically increased interest in charismatic leadership and for the Technocrats' success in passing themselves off as visionary. But this enthusiasm, although perfectly understandable, is misplaced.

What is vision, anyway? If we all walked around with binoculars affixed to our eyes, would life be better? We would see farther, you say. Yes, but we would stumble on the underbrush of life. Myopic people are obliged to use glasses to see, let us say, midrange objects. I remember, at age six, trying on my best friend's glasses and exclaiming, "I didn't know you were supposed to be able to see the leaves on the trees!" To myopic me, the trees were just big green balls. But, if farsighted people use my glasses, they can't see anything clearly; they lose their midrange sight. If I use reading glasses, I lose my short-range sight. And so on and so on. What's the point of this seemingly pointless digression? Well, the fact is, we, and organizations, need all kinds of vision—short-, medium-, and long-range.

Will it come as any surprise if I say that all organizations need Artists at some point in time, at the top or in some specific function like R&D or marketing or finance? Artists are the ones with the binoculars. If you can get your hands on a real one, so much the better. But your odds are about one in a hundred, and Artists are tough to recognize. By their deeds are they known, not by their talk. Of course we need visionary leaders, but there's some truth, too, in the Craftsman's comment that "He [Cobb] never implemented the head-office systems that would have allowed the decentralized management of the subsidiaries. So, the wheels started falling off all the carts." Wheels tend to fall off carts if you don't pay attention to them from time to time. Artists have a tendency to ignore the details. It's not that they're not analytical—they are; James and Cobb could perform imaginative financial contortions on a scrap of paper that would make your head swim. But don't ask them to work out all the minutiae. They're not interested in the details. They get impatient with them. That's why they delegate so easily: "Here. Here's the outline. Run with it. I have to see so and so, go here, go there." Details bore them. That's one of the reasons they surround themselves with Craftsmen and Technocrats.

The Craftsmen, on the other hand, respond to the Artists' visions with things like, "Okay. I like it but you can't get there from here. It's too great a leap. We will have to do such and such by 1997 if you want to get there by 2000." They're practical, realistic. Rome can't be built in a day. Systems and people can't be changed overnight; one told us, "It takes time to weld a team together and to solve human problems. They're not machines that you can just turn on and off." Most people need stability more than they need, or can handle, change—you know, like those stress scales that tell you that if you've changed jobs, cities, spouses, whether happily or not, your stress level is likely to be very high. Change, good or bad, is stressful and disorienting. The Craftsmen know and respect this. They're wise and realistic, and their vision is medium-range.

There's another kind of vision for which we have expressions like, "He went over it with a fine-tooth comb," meaning he examined it in great detail. We want our lawyers, for example, to examine contracts very carefully. We get them to read the fine print. We're not happy at all if they say things like, "It looks good to me," in a cavalier way. I don't know about you, but I want my airline pilot to look very carefully at all the gauges and dials before we take off. So, more generally, we need a kind of microscopic vision too. Guess who's good at microscopic vision? Yep. The Technocrats. They'll tell you, gladly, all the things you're doing wrong, all the short-range weaknesses in the system that may impede, or even render impossible, the accomplishment of medium- and long-term goals. This is necessary; as Benjamin Franklin reminded us, "*A little neglect can breed mischief . . . for want of a nail, the shoe was lost; for want of a shoe the horse was lost; and for want of a horse the rider was lost.*" Then, although the general had a great strategy, he lost the battle. The trick is to keep this short-range view in its place and under control. I don't want to throw analytical brilliance out. I want it working for me. I want it subservient to *humane* and *realistic.* The Technocrats are great at managing things; I just don't want them to have any authority over people or dreams.

Does this mean we can dispense with experimentation and with planning if we have vision, if we have Artists? Or dispense with vision if we have experimentation? Or dispense with plans and control if we have both? I don't think so. You see, these theories are not wrong; they simply try to do too much, to cover too much ground, to be all things to all people, to all times, and to all organizations. They leave out context, particular tasks, particular times, particular organizations.

The planning mode is right for organizations that already know their mission, not for finding it. I read recently that it takes General Motors eight more hours to assemble a car than it does Ford.[4] It seems pretty clear to me that GM needs a plan. It may also need a vision and a sense of experimentation, but that's another matter.

The experimental mode is right for organizations to adapt their mission to day-to-day, month-to-month realities. There's a great deal of experimentation going on these days in the financial services industry as people jockey for position and as consumer needs change with the bulging, now middle-aged baby boomers. Does that mean that financial institutions don't need a plan or a dream? I don't think so.

The visionary mode is necessary for organizations that have lost their way, but it is false to say that all our organizations and institutions have lost their way. (Chairman Mao believed in permanent revolution and look what it produced.) James's initial vision sustained an organization for at least 20 years. Who knows? The two other Artists, Mike and Cobb, might have sustained it for the next 20 years had they been given time. But no, we pull the rug out from under them as soon as they seem to falter. We get scared. Anyway, the visionary mode might be necessary in certain functions at certain times. For example, it might be good in marketing, in R&D, and even in finance; creative financing has allowed many a company to pursue its long-range dreams. In governments, it seems to me we need some vision right now in the policy-planning apparatus, but not necessarily at the top at all times. *It depends.*

Finally, these theories about organizational mission are incomplete, because they leave out character and its influence on judgment. With what we now know about the Artists, the Craftsmen, and the Technocrats, which mode do you think each will prefer? Don't you think that the Technocrats, with their need for security and order, will operate by plan and defend to the death that this plan is absolutely the only way to go? Experiments sometimes fail, and we know how they feel about "mistakes," and we know how their employees work to rule to avoid making mistakes, like the Craftsman who told me, "I'm trying to turn myself into a robot, but it's not working very well." And, we know how the Technocrats feel about vision; they think visionaries are dreamers, confused, not serious. They don't keep their noses to the grindstone.

Is it any wonder, then, that the idea of strategic planning held sway in the past 20 years, since it was so well suited to the technocratic mentality that had triumphed?

And the Artists. How do you think they feel about plans? One told us how: "That stuff's just an excuse not to work. All they do is produce tons of paper that just goes into a shredder eventually anyway, so what's the point?" And a Craftsman adds, "I just keep my head down and look after business." They don't like plans either. For them, plans are unreal. They like experimentation. They like trial and error, taking small risks. They like getting better and better at what they already know how to do. They'll use their old tool first, but try a new tool to see if it performs better on an exotic piece of wood with which they're unfamiliar. They'll seek advice from someone who's already worked with it. They keep an ear to the ground for information.

So, I conclude, these theories are not, I repeat, wrong. They seem to advocate mutually exclusive options—planning, experimenting, imagining—which are not mutually exclusive but which depend heavily on the context, on the task, and on who is in the driver's seat. Efficacy, doing the right thing, means doing the right things today, next week, and next year. So, I claim, organizations need all sorts of vision, short-, medium-, and long-term. One is not a substitute for the other. The trick is to have the right people in the right place at the right time and to make sure that the Technocrat has loads of influence and not a shred of power.

Expertise: On Getting Better and Better at What We Do

How do we acquire expertise? With experience, with learning. The Learning Organization has become a new catchphrase, but as I've tried to argue, organizations don't learn, people do. So let's look first at how people learn and then ask ourselves what that means for organizations.

I claim that there are essentially three ways in which people, and humanity (which is just people writ large), learn. Santayana told us, "Man's progress has a poetic phase in which he imagines a world." The poetic phase, or, if you like, the poetic *way* of learning, is discontinuous. It breaks radically with the standard way of viewing things. It unlearns in order to learn—"A painting moves forward by a series of destructions; what would live in song immortally, must in life first perish." This form of learning is deeply personal; it stems from the peculiar vision of an individual. It is not a collective enterprise. Culturally, we have always relied on our visionaries to lead the way. In science, we call the visionary a *genius,* like Einstein; in letters, a *poet,* like T. S. Eliot; in politics, a *statesman,* like Churchill; in business, a *leader;* generically, an *Artist.* But the Artist, far from ignoring the past, must possess it in order to break with it. The imaginative leap comes to the prepared mind (like Copernicus or Einstein), and the unprepared mind, without experience, is just an amateur, a dabbler.

Then there is continuous learning. The new insight offered by an individual genius is transformed into a collective enterprise. We take the insight, which is usually fragmentary and vague, and we worry away at it. We work out the bugs. We put flesh on the skeleton. The essential method of this continuous learning is trial and error; you work on my error with another trial. We try to articulate, to make clear and concrete, the new insight. Kuhn calls this *normal science* in the domain of science, but it is true in all domains. James's insight, his dream, his metaphor, was sculpted, shaped, made concrete by the skilled Craftsmen he trusted and to whom he granted authority. Without experience, without knowledge of the past—what has worked and what hasn't—there is no learning.

Finally, there is a third form of learning, and it requires neither imagination nor practice to make it our own. It requires diligence and concentration, not skill or imagination. It is called *reading, studying, analyzing, examining other people's work.* It, too, is largely an individual endeavor and not a collective one. We call someone knowledgeable who performs this well; if, after a great deal of analytic (not life) experience and diligence, he or she performs it very well, we call them brilliant.

So experience is essential to imagination, to skill, and to brilliance, to all three forms of learning. I see this every day in a classroom. In any given group of 30 or 40 students, there are always one or two who don't need me; their insights are stronger and more peculiar than my own—Artists. There are always two or three who are neither diligent nor imaginative nor skilled and cannot (or will not) learn—fools. And there are always one or two who, rigid and arrogant, claim to know everything already—"It says on page 24 that that's how you do it"—Technocrats. Then there is the great majority, varying in native intelligence to be sure, but all capable and willing to learn collectively, from each other and from teachers like me—Craftsmen. We also need to learn from books, from magnificent lecturers, and from the sudden inspirations of vision.

So, in an organizational context, we need people who have different ways of learning. The North American auto industry, it's alleged, foundered on the rocks of Japanese quality. We hear talk about creating higher quality. Well, where does quality come from? It depends. Can't it come from paying meticulous attention to detail? Can't it come from diligence? Producing an automobile, for example, requires producing a large array of parts, each one perfect. You don't want to receive your new car and have the handles fall off the doors—"Oops. Sorry. We missed that one."

Can't it, doesn't it, also come from trial and error, from experience? How can you know that something new will work if you don't know what has already been tried? Aren't all the automobile companies learning from one another, copying one another, and making marginal improvements on one another?

Of course, we also learn from fundamental innovation, which reshapes the way we look at the world. Like the way the Mustang reshaped a segment of the market. Like the way Macintosh reshaped the image of the computer. All the studies of major innovation show that it comes largely from individuals, not from groups or from large organizations. For that we need inventors.

Will it come as any surprise, then, if I say that organizations need all kinds of people who learn in different ways?

Efficiency: On Control

Since Fayol's day in the early part of this century, how to organize to accomplish tasks efficiently, with the least amount of effort, has been a serious preoccupation in management. Until, I suppose, the 60s, efficient organizing was seen to be pretty straightforward and was based, essentially, on *span of control.* People quibbled over whether one ought to have five, six, seven, or eight people reporting to any one manager, but few quarreled with the idea of span of control as a basic organizing principle. Span of control, of course, presupposes authority and hierarchy, and, as of at least the 60s, both of these principles have been undermined. Early on, they were contested by things such as "matrix management" (read: many different bosses), later by "self-organizing organizations" (whatever that means), but most frequently by something more general called *participative management.* Now, participative management means many different things to many different people; there is no single definition nor one all-important author or book on which we can rely. It's sort of just floating in the air—which makes me suspect that it may be rather light.

Who could be (publicly) against participation? I think the participative-management impulse represents an intelligent, normal reaction against authoritarianism. It's a reaction against the idea that grown people should be treated like children and told what to do. In fact, we are no longer even allowed to treat children like children, so I guess it stands to reason we shouldn't treat adults like children. Or does it? We have been told that authority structures stamp out creativity in both adults and children; that organizations need the creativity of all their members; and that, therefore, authority structures are bad, bad, bad. But are they?

No one likes authoritarianism, but *authoritarianism,* as my friend and colleague Laurent Lapierre has reminded me, is not the same thing as *authority.*[5] Authority has to do with setting boundaries and limits. As New York Mayor Rudolph Giuliani stressed in a recent speech, freedom is about authority: "We see only the oppressive side of authority . . . [but] freedom is about the willingness of every single human being to cede to lawful authority a great deal of discretion about what you do."[6] (But, if authority is seen as incompetent or illegitimate, the wish to participate [lynch], increases.) Great art, too, depends vitally on limits. If there are no limits, no authority to transcend, no traditions to fight, you don't get Artists; you get amateurs—necessity is the mother of invention. Limits are the reality principle in operation; the schizophrenic who produces paintings creates images that are interesting but meaningless.[7] The parent of every adolescent knows the importance of limits. Adolescent psychiatrists tell us the limits have to be there so that the child can define himself or herself against them. No limits, no identity. It's tough, and tougher and tougher all the time, to define such limits, but it must be done. (In parentheses, British child psychiatrist D. W. Winnicott says that the task of every parent of an adolescent is survival—just survival.)

There are some emergency situations, like war, that require strong authority. Let's listen in on a hypothetical conversation in the World War I trenches:

CAPTAIN: So men, I thought we might go up into no man's land tonight. What do you think?

SOLDIERS: Gee, I dunno captain, it's pretty bright tonight and . . .

Sometimes authority is indispensable.

Ah, but that's a false analogy, you say. You can't compare organizations to war. Okay, how about this? We once hired a gold-medal-winning MBA graduate to work for us. The person was relatively experienced and very bright. We hired her to develop a new aspect of our business. I told her, "Here it is. Here are the basic parameters. How it gets done is up to you." Every week or so I asked her how things were coming and she told me that she was writing up a plan. Good, I thought. We've finally got someone serious around here. Someone who knows what she's doing. (In those days, I was very impressed by plans.) Well, time wore on, and on, and on. People began to suggest to me, not very subtly, that she should be fired. Finally in exasperation I asked her what was wrong. She replied, in equal exasperation, that she didn't know what I wanted and so she couldn't do a plan. Some people, no matter how bright or well trained they are, can't fill vacuums; they need direction, and I'd been doing her no favor at all to ask her to work autonomously. She was miserable. It made her feel totally, but totally, inadequate. This was not kindness, it was inhumane. I gave her a few more details—the few I possessed—and afterward she performed beautifully and, sometimes, creatively.

Take another example. Suppose we are in a regime of participative management interpreted as management by consensus. We have four people who are bright and creative and hardworking, and one lazy son of a gun who contests everything everyone else wants to do and has a right to be heard. Doesn't listening to him constantly just demoralize everyone else? Is participative management letting everyone have equal airtime?

Take another example. Between the two great wars, people wanted to play. In England, everyone thought that the country's military preparedness was perfectly adequate. Churchill disagreed, and it was Churchill and not the majority who was right. During the same period in France, de Gaulle was right. Does participative management mean rule by the majority? Does it make leadership unnecessary? Will someone please clarify this for me? Does it just mean listening?

If, to take the best-case scenario, it means listening, who is good at it? Do the Artists listen? Well, yes they do, if you can catch them between airplanes. Do the Technocrats listen? Nope. If you're looking forward to participative management as listening, you're in for a big disappointment with them—although they talk a lot about listening when masquerading as Craftsmen. Who listens attentively all the time? You got it. It's the Craftsmen. Do they abdicate authority? Never. Do they take responsibility? Always—even if the bad idea came from one of their apprentices.

While we have been led to believe that control and participation exist along a continuum with control at one end and participation at the other, I look at this participation thing in the following way (see figure).

It depends. It depends on what you want to accomplish and how urgently. In the university, for example, change takes a very, very long time and is hideously inefficient by design. We wish to preserve the value of participation, we seek consensus, and academics don't willingly give up control. Society as a whole gets very

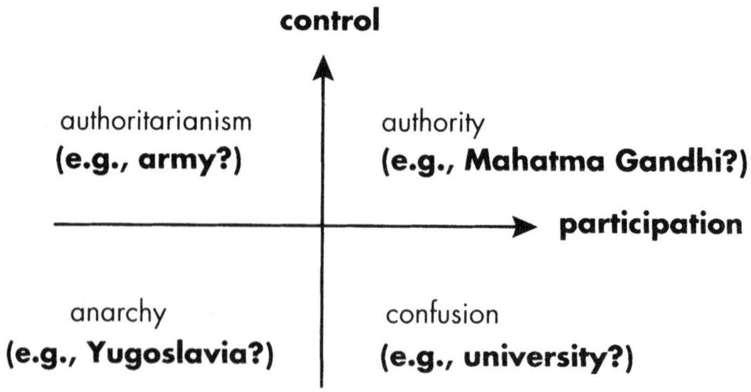

frustrated by our inability to respond instantly to its desires. (I, for one, don't think universities should respond quickly to society's whims. That's not our job.)

The former Yugoslavia is a current example of anarchy, with no control and no formal, institutional, structured way for people to participate in the events that shape—and destroy—their lives. Anarchy is inefficient both because of an absence of control and because of an absence of structured participation.

The army, I submit, is the place for authoritarianism and strict chain of command, and it would be inefficient with more participation. So, I argue, would the cockpit of an airplane.

PILOT: Joe, would you check the fuel gauge, please.

COPILOT: It's not my job.

PILOT: I've got my hands full here.

PASSENGER: Would *someone* please check it!!

In short, there's a continuous, delicate tension between the demands of control and the demands of participation. It can't be settled for all time, and not all organizations need more participative management. In fact, participation can sometimes be abdication and sometimes a smokescreen: You turn a debate over to a group which you know in advance will not be able to reach consensus, and then you step in like an "objective" third party and say, "You see. These people are too emotional to decide anything. Leave it to me." Management is, I contend, a judicious and timely blend of control and participation. When control, when participation. I pick Mahatma Gandhi as one of my heroes. He had a vision, and a sometimes domineering personal authority, but he also knew how and when to listen. You, no doubt, have your own example.

In general, efficacy, expertise, and efficiency all depend on what you want to accomplish and on who's doing the accomplishing. We have to reject all-purpose recipes. All-purpose flour is a great boon to ordinary cooks, and I may win a local pie contest using it. I have no illusions, however, that I would win in a Parisian culinary contest. Most of us don't even want to enter such a contest.

Most of us, and most organizations, just have tasks to accomplish that sometimes demand a lot of participation and sometimes a little; that sometimes demand a lot of authority (not authoritarianism) and sometimes a little; that sometimes demand a wholly new vision and sometimes just demand a lot of hard work. Some organizations do a lousy job of learning because they are packed, as our example was, only with diligence and not with imagination or with skill and experience. There were no new visions because there were no Artists, just trend-line projections into an uncertain future. No experimentation, because there were no experimenters; no quality, because there was no *cult of excellence*—in short, no Craftsmen. People, not systems and recipes, make the difference.

How could we possibly have forgotten this? Because, for a variety of reasons, we have forgotten character. We are the slaves of some defunct philosophers. We have been taught that behavior is everything: behaviorism. We have been taught that thinking is everything: human as information processor. We have been taught that all people are created equal: egalitarianism. They are created *equal* in the sense that they deserve respect, but they are not created the *same.* These three things, and no doubt many more, have led us to imagine that anybody can be taught anything if given half a chance and the right method. We imagine that you can just lasso *acting* and reform behavior with teaching—tell people how to be inspiring. Or you could just lasso thinking and teach people how to be visionary by thinking sideways or upside down. Or you could lasso *seeing* and teach people how their perception is distorted, so that there'll be no more, say, racism. But character doesn't work that way. It's a package. You can teach Artists to watch for the details, but they can't walk the talk. You can teach Technocrats the language of empowerment, but they

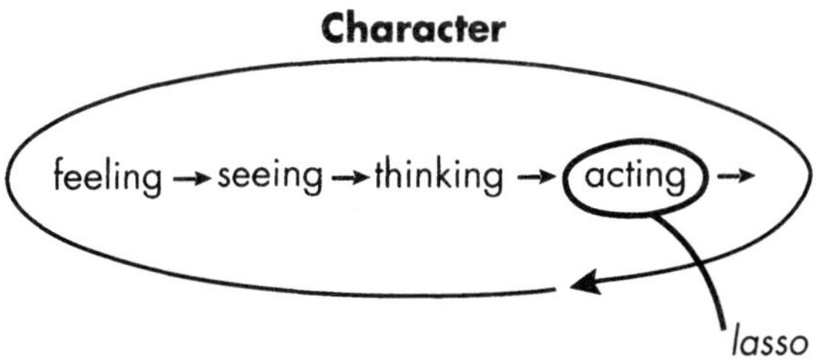

can't walk the talk. You can teach Craftsmen the language of vision, but they can't walk the talk.

So, that means you need diversity in the management team: people who are good at inspiring loyalty and participation; people who are good at dreaming; people who are good at the details. The trick is to get them to work together in the same direction. How do you get teamwork? Certainly not by pious and hollow phrases like "respect each other," "communicate with each other." Do Artists work well in teams? Well, yes, in the sense that usually they know they need somebody else to complete their strengths. Do Craftsmen work in teams? Well, yes, most of the time. Even Technocrats work in teams providing they're not at the top, providing they don't have power. So, what's the biggest barrier to teamwork? Technocrats in positions of authority. Why? Because they don't respect anybody else, and they don't think they need help, so how are you going to get them to come to a seminar on teamwork (except to look good to the boss)? All right, let's see what the scope of the problem is. How many are there in positions of power?

12

The Scope of the Problem

Things fall apart; the centre cannot hold;
Mere anarchy is loosed upon the world,
The blood-dimmed tide is loosed, and everywhere
The ceremony of innocence is drowned;
The best lack all conviction, while the worst
Are full of passionate intensity.

W. B. Yeats, *The Second Coming*[1]

No one knows the scope of the problem, but we all sense its importance. We can only speculate. Here are my speculations. You can, of course, make your own.

First of all, I think there are signs all around us of a profound social and economic and political malaise. People are variously angry, frustrated, depressed, violent, withdrawn, radicalized, or in despair. *"The best lack all conviction, while the worst are full of passionate intensity."* This is the social malaise. The economic malaise is felt in a loss of competitiveness and in the faltering of prominent economic institutions like IBM. The political malaise is evident in our lack of trust in our institutions and our politicians, low voter turnout and, worse, Oklahoma City. "*Something is rotten in the state of Denmark*" (Hamlet); something is going badly wrong and it's not just in business. What has gone on in business is just a reflection of a much larger social phenomenon, the Triumph of the Technocrat, and changing our business leaders won't be enough to make things right again. It will solve some economic problems, but the social and political ones won't go away.

Here is my estimate of the scope of the problem. I hasten to add that this is not "scientific"; it's impressionistic. As I said, it feels like the Technocrats have taken over the whole of life. But if that were true, most Western economies would be in total collapse. So, realistically, how many Technocrats are there?

General Population

In my experience with both the private and public sectors, Craftsmen represent the vast majority of people in the general population at large and in most subgroups. In my batch of university students, for example, whether they're studying economics, human relations, or accounting, the majority are Craftsmen. Both Artists and Technocrats are in a distinct minority. For the sake of illustration, to give ourselves an image we can think with, let's use the famous normal curve known as the *Bell Curve.* In Figure 12.1 we represent the majority under the main part of the curve and let's say it represents 80%. Under each tail we have Artists and Technocrats each occupying about 10%.

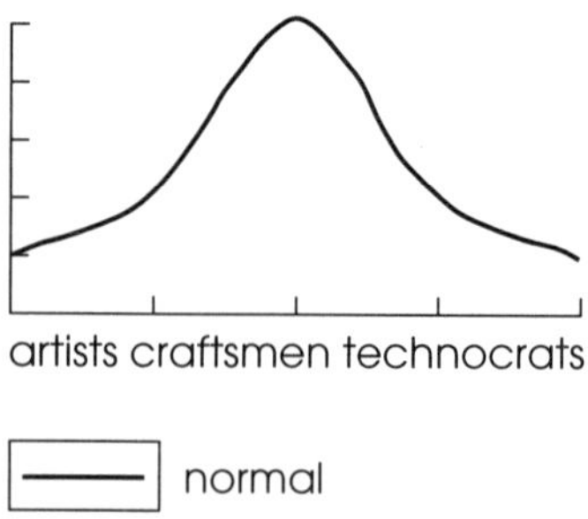

Figure 12.1 General population.

In Big Business

A friend of mine who runs an investment bank and who deals with CEOs all the time on financing deals, estimates that about 40 percent of current CEOs are Technocrats. If he's right—and I have no reason to doubt him—then my conservative estimate is as follows. The further up the hierarchy one goes, the more the craft population dwindles and the more the other two increase in importance. By the time you get into the upper echelons of most organizations, we have extreme underrepresentation of Craftsmen and overrepresentation of both Artists and Technocrats in proportions of one-third each. Most of the artists visible in big business started a small business and ended up a big success—people like Perot, Jobs, Gates, and Turner. Few Artists manage to come up through the ranks of large business.

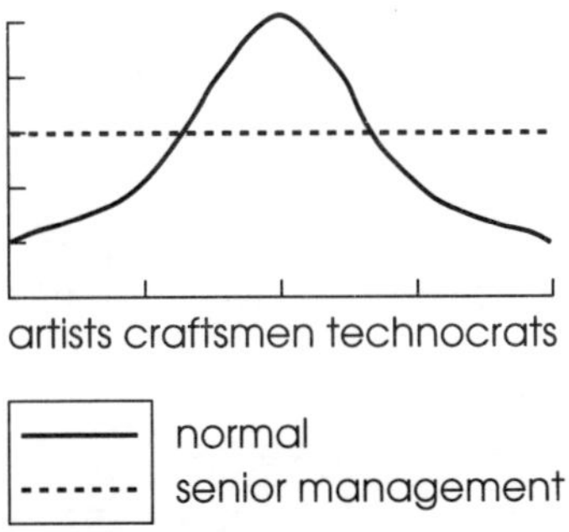

Figure 12.2 Big business.

Other Countries, Other Cultures

But that depends, in part, on what country you are looking at. Figure 12.2 describes what I think the ranks of American senior managers look like, taking into account a very high level of new business formation. All countries do not benefit from America's high level of entrepreneurship, nor do they have America's broad and deep capital markets to finance it. If we look at just two other examples, France and Germany, I think the portrait may be quite different (see Figure 12.3). Well, let's see why that might be the case. Take Germany. The German culture values craftsmanship and apprenticeship much more than does our own. It's a very rare German CEO who did not come up through the ranks of the industry or the firm and who is not well over 50 before reaching the top slot. In-house experience counts. In a study[2] of the top 200 CEOs in Germany and France, we learn that 32 percent of German CEOs came from inside the company. The comparable number in France was 4 percent, where a full 36 percent were recruited from the public service! What the American economy has lost in the technocratic takeover, it has made up by its entrepreneurs. The German economy has stayed strong because of its emphasis on skills and experience. The French economy is weak, in part, because of a lower level of entrepreneurship and because too many large French companies may be run by Technocrats who have no background in the company or in the sector. Brains alone aren't enough to run an organization.

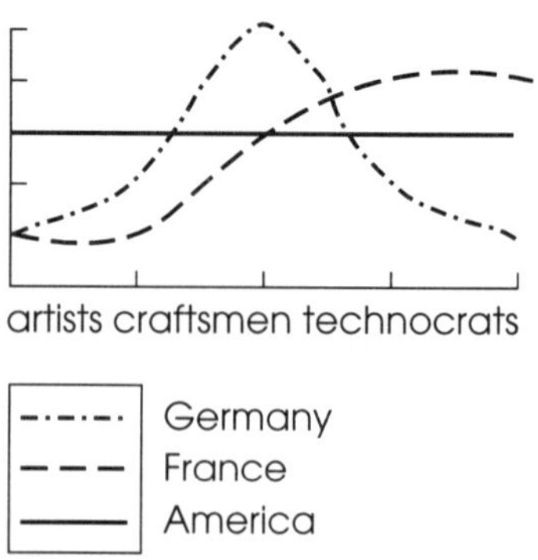

Figure 12.3 Senior management in other countries.

The Public Sector

And that brings me to the public sector. This is really scary. In the private sector, the market disciplines bad management. You go bankrupt. In the public sector, there is no disciplining mechanism. Responsibility is too divided to know who to blame. It's always some other department's fault. The only thing citizens know is that it's not working. Technocrats are attracted to government like flies because they look for power and prestige. They flatter their superiors and get ahead easily. And, you can't fire them. So, the curve looks like Figure 12.4, heavily weighted toward the Technocrats. The Craftsmen, people who genuinely believe in public service, have precious little power. And, as for Artists, well . . .

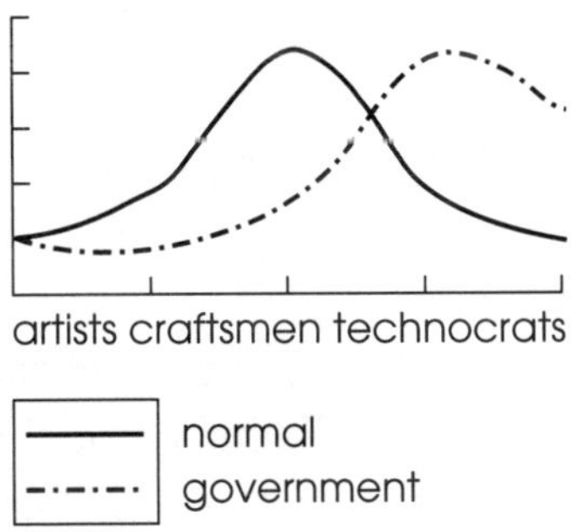

Figure 12.4 Senior public sector.

In Small Business

The curve in Figure 12.5 looks vastly different (which may help to explain why small business and governments are natural enemies). Businesses are started by Craftsmen and Artists. The really big success stories are invariably Artists. Technocrats are often brought in later on to run things and often destroy what was built. Technocrats haven't got enough imagination to start something,

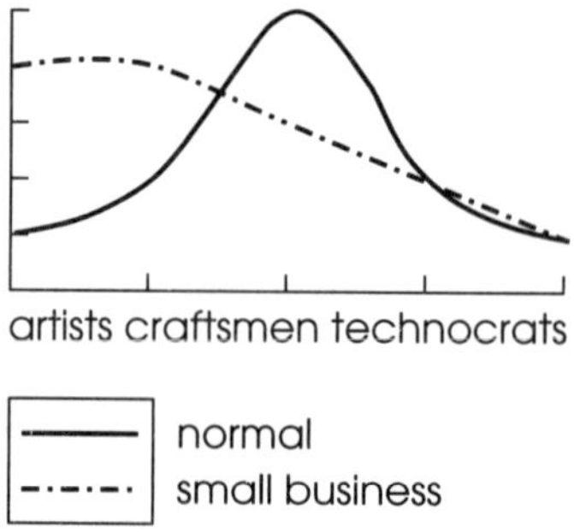

Figure 12.5 Small business.

they're extremely risk-averse, and there's not enough prestige involved in small business. Too much hard work and not enough glory. There is one exception to this rule. After a speech I made about this recently, a woman came up to me and said, "I think you're wrong about Technocrats not starting businesses." "Why?" I asked. "They start consulting firms," she said. I had to laugh because, of course, she's right. Not that all consultants are Technocrats. On the contrary, many consultants are Craftsmen who have been fired by Technocrats. Often these same Craftsmen turn up later as consultants to the original organization that's trying to put itself back together after a Technocratic flu. Many consultants are Artists who are solitary and uncomfortable working for someone else. But consultancy is fertile terrain for Technocrats, because you get to play God. You get to apply all-purpose recipes and rules to somebody else's business, with somebody else's money. No risks, no responsibilities, and high pay. Perfect.

Women in Power

Why should I single out women? Because there's a widespread myth that we are in all this trouble because it's a man's world. If women were in power, so the myth goes, everything would be just hunky-dory. Not so. Women are not all sweet, cuddly, humane Craftsmen or creative Artists. In my experience, women who manage to make it to the top of major organizations—and stay there—are some of the worst Technocrats of all: cold, crisp, cerebral, dressed up like men, or flaunting the outward signs of femininity in clothes or makeup to compensate for deficiencies on the inside. Whether because they've had to be better, more technocratic, than the men to get ahead or because they were very technocratic people to begin with, and therefore got ahead, who cares? They're there (see Figure 12.6). So, choosing a woman to run your organization is no guarantee of anything.

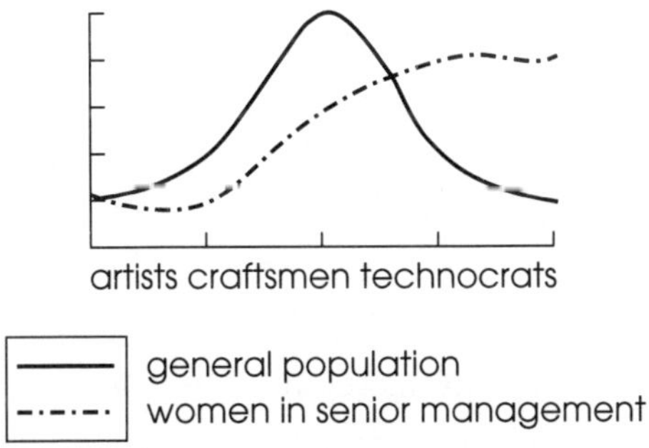

Figure 12.6 Women in power.

Truth or Consequences

The technocratic mentality frowns on the participation of "imbeciles." It centralizes power, it disenfranchises ordinary people in favor of "Experts," and, by a curious twist of democratic logic, we get government "of the Experts, for the Experts, and by the Experts." Public institutions are too big, too remote, too detached from reality and are governed by theoreticians and rule books.

Teachers feel they have no say in what happens in the classroom. Doctors and nurses feel they have no control over health care. Employees feel they are not consulted. Voters feel that they have to choose between Tweedledum and Tweedledee. Many stop bothering to vote. We don't trust anybody.

Collectively, we have been driven to fight fire with fire. Faced with technocratic corporate leadership, unions elect themselves technocratic leaders to fight back. Faced with technocratic rule books, school administrators hire Technocrats to fight back. Faced with technocratic resistance to environmental issues, environmental groups hire rigid, dogmatic, cerebral, analytical, closed-minded Technocrats to fight their battles. Faced with technocratic male resistance to change, women's groups became technocratic. Fight fire with fire. Except that now we have a conflagration.

Technocrats come in all shapes and sizes and all guises. They are in all sectors—not just in private organizations, but in government, in nonprofit associations, in politics, in movements as advocates of all sorts of different causes. You can't identify them by the causes they espouse, but only by the *way* they espouse them. We need to develop an instinctive capacity to recognize them. Whenever I meet someone who makes me feel that I am (*a*) a woman (emotional), (*b*) a child (immature), (*c*) an imbecile (stupid), or (*d*) all of the above, I know I'm dealing with one. Recognition starts in *feeling* and proceeds to thought and judgment; as seventeenth-century French philosopher Blaise Pascal said, "The heart has its reasons which Reason does not know." I am now able to sense a Technocrat at fifty yards. I start to feel uncomfortable, slightly irritated, slightly suffocated. Gradually I'm able to put thoughts and names to my feelings. I hear a little sermonette from him and I know he thinks that everyone else is an imbecile, that only he possesses Truth, Justice, and Morality.

She seems brilliant, but I sense rigidity. I know her mind is closed. I think we should start to develop a healthy suspicion of the description *brilliant.* We used to have expressions that tried to capture this suspicion. We said, "Oh, him? He's too clever by half." The expression has fallen into disuse. "Too clever by half." What did that mean? It meant too clever, not for his own good, but for ours. We used to mistrust cleverness. "Oh, yes, he was a clever fellow," meaning, he was a fancy talker, too clever for us. "What a clever child!" meaning ingenious and sometimes meaning, if you can't say something nice don't say anything at all; that is, the person wanted to say "what a brat" but found something nice to say instead. The earliest etymological origins of the word *clever* associate it with *claw.*[3] That ought to tell us something! *Brilliant* and *clever* are both purely associated with the head; they tell us nothing about the heart, about imagination, about ethics. We should learn to say, "Brilliant, huh? That's nice. What else have you got?" And we should watch and listen for quite a while before we buy, because *brilliant* can be associated with either the Artist or the Technocrat. We should become unrepentant doubting Thomases or, if you like, we should all "come from Missouri."

She's uncompromising and hardheaded. These two adjectives capture what I would put under the heading *disdain,* or disrespect. I think we need to develop antennae to detect disdain. The technocratic mentality is very sure of itself. Let's listen in on some hypothetical conversations:

> TECHNOCRAT: And so, in conclusion, the only crisis that matters right now is the ozone layer. We must shut down factory X.
>
> A MEMBER OF THE AUDIENCE: I was just wondering what we will do with all the people who depend on that factory for their livelihood. And what about the town?
>
> TECHNOCRAT: You've fallen for the same old line of reasoning. We can't worry about that; the facts show that the whole planet is at risk. The facts speak for themselves (*you sentimental imbecile*).

Another context and another subject:

> TECHNOCRAT: And so, in conclusion, we have to ban smoking from the planet.
>
> A MEMBER OF THE AUDIENCE: I was reading the other day in a medical journal that major depression and the propensity to cigarette smoking are highly correlated. These doctors were suggesting that we should be a bit careful with depressed people, because pushing them too hard to quit smoking may inadvertently push them into a major clinical depression.[4]
>
> TECHNOCRAT: The study was no doubt financed by the cigarette lobby [paranoid], and even if it's true, that's their problem [narcissistic].

Another context and another subject:

> TECHNOCRAT: And so, in conclusion, we have to lay off fourteen hundred and thirty-six people in the widgets division.
>
> A MEMBER OF THE AUDIENCE: I was just wondering if we couldn't use some of those people to develop Edward's new product idea.
>
> TECHNOCRAT: Ed's new product idea will never see the light of day. Ed's a nice guy but he's a dreamer. Instant photography. Honestly. It doesn't make sense. It's not *logical.*

Now, the Technocrats might be right about the ozone layer, about smoking, about photography. Because they are brilliant, they often are right, but that's not the point. The point is that they're absolutely certain they're right; they're never wrong and other people are just too stupid to see it. The Technocrats never doubt, never get depressed, always have an answer. Lucky them! Most of us live with a crushing sense of doubt. So, again, watch them. Does he ever waver? Does he ever question himself? Does she ever mock herself? Does his certitude ever falter? In short, beware of unflagging certitude in a messy world, and beware of anyone who just parrots conventional wisdom of whatever ilk, whatever political or social stripe.

If this is the scope of the problem, how do we get out? In the next chapter, I sketch out what I think to be part of the elements of a solution. Not the whole solution. Not a final solution. Not a miraculous solution. Not a quick solution. It starts by laying out how we got here. If the diagnosis is wrong, the treatment won't usually do much good.

13

Partial Solutions

Practical men, who believe themselves to be quite exempt from any intellectual influences, are usually the slaves of some defunct economist.

John Maynard Keynes

Or some defunct philosophers. Some analysts blame a bunch of dead Frenchmen; we have popular books with titles like *Voltaire's Bastards*[1] and *Descartes' Error.*[2] More thorough philosophers[3] trace it to a long tradition running through Descartes and Voltaire, to be sure, but also through the British bloodlines of Locke and Hume and a host of more modern philosophers. The Triumph of the Technocrat has many philosophical fathers; *success has many fathers, failure is an orphan.* The Technocrats are the consummate Enlightenment Man taken to his ridiculous extreme. They are the inevitable progeny, the caricature, of the triumph of "instrumental reason," "disengaged reason," dispassionate logic, or the dominant idea that reason divorced from emotion is the guarantor of sanity, "objectivity," and good judgment. According to legend, the person with the least emotion is, necessarily, the best decision maker.

The Technocrats thrive and prosper because of this false idea. In a sense, their triumph has a Darwinian flavor about it; just as someone seven feet tall may be good at basketball, theirs was simply the species best adapted to the social and intellectual climate of the post–World War II world. It was a world that believed in inevitable technical progress steered by enlightened social and economic engineers. Dispassionate, altruistic men and women who, by their force of intellect, could plan and manage whole societies, indeed, the whole world. The more brilliant and dispassionate, the better.

Except that it didn't turn out that way. The various social and economic and political breakdowns we are experiencing challenge the very basis of the triumph. There is a flight from high Reason into cults, celestial prophesies, fanaticism, and violence. Instrumental reason, utilitarian reason, cold objectivity, Reason itself, are being rejected in favor of "passion," "creativity," anarchy. Another baby being thrown out with the bathwater. But Reason is not the enemy; in fact, Reason, as usual, will be our salvation. How? Well, we're going to *think* it through and *feel* it through—reuniting what has been torn asunder. Reason and feeling, thought and emotion. And, we have some *scientific,* as opposed to *prophetic,* help.

Seven-foot-tall Technocrats are not, on the whole, evil. They are, on the whole, well-meaning. But the road to hell is often paved with good intentions. As individuals, they are not the enemy; as a caste, as a governing class, they are. And not because it's a conspiracy. It's not. There is no international plot. There is no plot at all. We are seduced by them because we are the slaves of some defunct philosophers. Technocrats are not, I repeat, evil. They are immature. They are handicapped. They are blind. Led by the blind, is it any wonder, then, that we are lost?

Technocrats and the Phineas Gage Syndrome

Why are they blind? Science to the rescue. Dr. Antonio Damasio, an American neurologist, has written the management book of the century. Its title is *Descartes' Error: Emotion, Reason, and the Human Brain.* Of course, he doesn't know he's written a management book; he thought he was just writing a scientific book about the brain. But it goes way beyond that. It's an intellectual watershed. Anyone interested in anything should read it. I can't do justice to it in a brief space, but let's have a look at highlights of the neurological evidence and at the central message.

Damasio opens his book with Phineas Gage:

> It is the summer of 1848. We are in New England. Phineas P. Gage, twenty-five years old, construction foreman, is about to go from riches to rags. A century and a half later his downfall will still be quite meaningful.

Phineas Gage, Damasio tells us, worked for a railroad, and his job was to supervise blasting to clear the way for laying down tracks. Because of a moment of distraction, a blast went off prematurely and drove an iron bar through his head: an iron bar with a diameter of 1¼ inches, weighing 13¼ pounds, 3 feet, 7 inches long. In other words, a *big* iron bar. The force of the blast was such that, after traveling through Phineas' head, the bar landed a hundred feet away. Phineas got up, was helped into an ox cart, and traveled about half an hour to a local inn, where he sat on the porch for approximately an hour recounting the incident to passers-by. He felt no pain and didn't seem distressed by what had happened. He was perfectly lucid and rational as he gave the attending physician the details. Within two months, he was pronounced fully recovered with all of his reasoning faculties intact. He never lost the capacity for speech, analysis, or reasoning. But, he lost his judgment. And his life went from bad to worse to total disaster. Why?

To the best of my understanding (I'm no brain surgeon), he lost his judgment because he lost his emotions, or the connection between his feelings and his reasoning. Central message? While we have been laboring under the presumption that emotions can cloud judgment (and obviously they can and do), we now learn that the *absence* of emotion can cloud judgment, too. If this sounds paradoxical, it is. Dr. Damasio tells us why:

> If you are wondering how bizarre it is that biological drives and emotion may be *both* beneficial and pernicious, let me say that this would not be the only instance in biology in which a given factor or mechanism may be negative or positive according to the circumstances. We all know that nitric oxide is toxic. It can pollute the air and poison the blood. Yet this same gas functions as a neurotransmitter, sending signals between the nerve cells. An even subtler example is glutamate, another neurotransmitter. Glutamate is ubiquitous in the brain, where it is used by one nerve cell to excite another. Yet when nerve cells are damaged, as in a stroke, they release excessive glutamate into the surrounding spaces, and thus cause overexcitation and eventually death of the innocent and healthy nerve cells in the vicinity. (p. 194; emphasis in the original)

Emotions precede and guide judgment. In the beginning, there is the body. According to Damasio's research, and that of his colleagues, emotional circuitry is laid down genetically at birth. Long before the higher cognitive functions develop, there is emotion. Hungry, the baby cries. Very gradually, through daily learning, the developing child begins to be able to reason about his or her feelings. But the thoughts are guided by, elaborated upon, a base of instinct and emotion. Negative emotional experiences will condition the thought processes and the reasoning attached to them. New channels, new neural circuits, are laid down between emotions and thoughts. Normal human beings are never without some background emotion. Human beings who are brain-damaged in the same areas as Phineas Gage are without the emotions, background feelings, that help reasoning and judgment along. In spite of otherwise intact cognitive functioning, they exhibit what Damasio calls a *myopia for the future;* they are unable to learn from past mistakes, and they therefore cannot make good judgments about tomorrow. They lose a sense of priorities and are too focused on the short-term. They *feel* no pain, so they have no signals to change a dangerous course. Normal people feel remorse, humiliation, guilt, chagrin when they make a mistake. If you feel nothing, why would you think you *made* a mistake? What would cause you to stop and reconsider? How could you learn?

In fact, the people we fear most in life are sociopaths and psychopaths: people who can kill in cold blood and are without remorse, without emotion. Damasio continues:

> Developmental sociopaths or psychopaths are well known to all of us from the daily news. They steal, they rape, they kill, they lie. They are often smart. The threshold at which their emotions kick in, when they do, is so high that they appear unflappable, and are, from their self-reports, unfeeling and uncaring. They are the very picture of the cold head we are told to keep in order to do the right thing . . . They are in fact yet another example of a pathological state in which a decline in rationality is accompanied by diminution or absence of feeling. It is certainly possible that developmental sociopathy arises from dysfunction within the same overall system which was impaired in Gage, at cortical or subcortical level. But rather than resulting from blunt macroscopic damage in adulthood, however, the impairment of developmental sociopaths would come from abnormal circuitry and abnormal chemical signalling and begin early in development. (p. 178)

Technocratic Parallels: In Cold Blood

Obviously, I am not saying that Technocrats are psychopaths, but I will argue that there is some missing connection, or some feeble connection, between their feelings and their thoughts that results in flawed judgment. Remember, they are very cold and cerebral. Let me try to clarify what I see as the parallels between this description and my understanding of Technocrats.

One Technocratic theme is "*What have you done for me today?*" They are disloyal, maybe in part because they have minimal doses of an emotion called *gratitude.* It doesn't matter to them that you may have worked all your life for the organization; what counts is today.

They have very little of another emotion called *empathy.* That may explain why they're so good at reengineering: They don't lose a wink of sleep by laying off hundreds of people. They can do it in cold blood.

What do they care about quality? Nothing. Why? Because of another emotion called *pride.* You know, that kind of warm, fuzzy feeling we get when were proud of ourselves, proud of the way we reacted in a difficult situation, proud of our humanitarian instincts, for example. Genuine pride is not to be confused with intellectual arrogance.

And how about *doubt?* In my experience, Technocrats don't have any. They are always sure of their calculations. If they had a little, just a little, doubt about their capacities they might surround themselves with people who think and behave differently from themselves. Just in case.

And how about *fear?* What if I make a mistake? What damage can I cause? What if I'm wrong?

And what of *remorse?* Robert McNamara has written a whole book billed as a mea culpa for Vietnam.[4] It's not a mea culpa, it's an apologia. He apologizes for not having got the *information systems* right. He apologizes for not having consulted enough *Experts.* Thirty years later, we can sense almost no *remorse* for the agonies through which he helped to put soldiers and a nation. The role that he played in the Vietnam drama might have driven another to the edge of a guilty suicidal depression.

To this day, my Technocrats don't feel *guilt* at the destruction of the dream. They blame the demise on bad systems, incompetent Craftsmen, idiotic Artists, the economy, anything and anyone but themselves. And the worst is, they're sincere. They believe themselves.

They can't learn from experience. They make the same mistakes over and over again. Some have gone on happily to destroy other organizations, sublimely oblivious to the damage they cause. Since they don't stay long enough to see the damage, they escape with reputations intact. Great work if you can get it.

They exhibit, to use Damasio's expression, a profound myopia for the future. All the while projecting artificial trend-lines into the future, they are totally concentrated on the present. Short term results. "Our strategy is profitability." Quarterly profits raised to a first principle.

They are blind. They are emotionally immature; like an adolescent, they think they possess the truth. They are handicapped; they have very bad judgment although they often have very high IQs. They are dangerous not because they are evil. They are dangerous because they are too cold-blooded. Just as we do not let children play with fire, we should not let emotionally immature people play with power. The Technocrats are no more to blame for their attitudes than are children for their immaturity. Unlike children, however, they can't be expected to grow up, and they can't be educated to change their ways. So, all we can do is ensure that they don't get access to power.

A Footnote on Education

Are MBAs all Technocrats? Are accountants? Are engineers and economists? In short, is education to blame? No, but it could be a partial antidote. First, a story.

Several colleagues and I recently wrote an article on decision making and sent it off to an academic journal for publication. I don't know if you're aware of how these things work, but the editors send such a submission out to other academics for review before they accept or reject it. In due course (18 months), we received those evaluations and the article was accepted. But that's not the story. The story is in one of the evaluations.

In our article, we had made reference to a *fifth column.* One reviewer objected that we had not shown the first four columns. I had an image of this studious individual's face reddening, his eyes bugging out, his hair standing on end, veins on his temples bulging, shouting, "Fifth column! Fifth column! Where are the first four?!!" But as you and I both know, *fifth column* is an expression, a piece of symbolism we owe to the military. In fact, its specific origin is the Spanish Civil War. Franco's fascists had four columns of troops marching on democratic Madrid and were said to have a "fifth column" of spies inside the city ready to rise up and destroy democracy from within. The expression has passed into ordinary language to denote spies, subversives, ideas or people that destroy from within.

How is it that someone with the highest academic degree that can be granted—a doctorate—a supposedly highly educated person, could not know that this was symbolic, that we didn't have to list the other four columns? How could such a person be functionally illiterate? Because he or she has probably never read a novel, or history, or philosophy, or sociology, or psychology, or even the newspapers. He or she has probably read nothing but management literature. *Yikes!* But if all you sociologists and psychologists think you're off the hook and I'm attacking business schools, you're dead wrong. It's a rare sociologist who knows something about psychology and a rare psychologist who knows something about philosophy. And nobody knows anything about the Bible, either as revelation or as literature. Too many of us are "experts," trained in specialties but *uneducated,* and the problem runs very, very deep. We are the exact opposite of the ideal we all profess to admire: the Renaissance men who knew science and letters and philosophy and art. What's worse, we allow experts to dictate policy to us. For example, French historian Alain-Gérard Slama reminds us that we allow cancer researchers who have studied biochemistry and medicine to dictate public policy on smoking, and yet smoking bans raise deep philosophical issues involving subtle tradeoffs between Justice and Liberty, which cancer researchers know nothing about.[5] Well, I'm no expert on liberal arts education, but Allan Bloom is, and anybody concerned about it should read *The Closing of the American Mind.*[6] It sure changed the way I look at my academic world.

In any event, the Technocrats thrive in an education system that encourages narrow specialization. In my experience, they're not very good at handling the subtleties of philosophy, history, or literature—subjects that don't easily reduce to formulas or one final answer. The more we insist on broad, general education, the less likely they are to come out at the head of the pack. Furthermore, being exposed to the world's best in science, literature, philosophy, economics, history, and mathematics is an inherently humbling experience. It's very hard to maintain one's self-assurance, to remain convinced that one has all the answers. And, even if a brilliant Technocrat manages to get through such an experience intact, there will be some domains where his or her colleagues are stronger. They will be able to fight back. But not, of course, if society doesn't value other arguments, if society as a whole remains mesmerized by cold-blooded analytical brilliance. So the real answer is to change us, change what we value. To do that we have to go back, not forward. As C. S. Lewis insists:

> Every age has its own outlook. It is specially good at seeing certain truths and specially liable to make certain mistakes. We all, therefore, need the books that will correct the characteristic mistakes of our own period. And that means, the old books.[7]

I am at a loss to understand why the wish to restore something of value should be seen to be regressive. Let's say you're driving down the road and you're thinking about your work. You're a bit distracted. You miss your regular turnoff and are forced to go back. Do people scream at you, "You ought to have kept going, you regressive, reactionary, antediluvian!"? Of course not. There's nothing to be ashamed of about retracing your steps if you've made a mistake. Maybe, as a society, we were distracted in the 60s and 70s and took a bunch of wrong turns. Maybe we need to retrace our steps. In any other context, this would pass for sanity, not reaction.

When you are driving in the fog, do you speed up? When you are lost, do you run? We have got to slow down, not speed up. We have to go backward as well as forward. Here's T. S. Eliot:

> We shall not cease from exploration
> And the end of all our exploring
> Will be to arrive where we started
> And know the place for the first time[8]

Why race headlong, like a car careering out of control, into a future that no one, no one, can know? It feels to me that that's what we're doing with our organizations and with society as a whole. Careering out of control. Rudderless. Pretending to know everything about the future and knowing nothing about the past. Everybody's reading the futurists and nobody's reading Santayana, and yet, he could have told us. The same Santayana who told us that "man's progress has a poetic phase in which he imagines the world" told us, if we had only listened to him in 1905, "Progress, far from consisting in change, depends on retentiveness. . . . Those who cannot remember the past are condemned to repeat it."

What do we need to remember? Craftsmanship. We have to get back to a sense of proportion and to the principles of Craftsmanship. Our world has been built by the slow accumulation of knowledge and skill, punctuated periodically by genuinely new—and, I must keep insisting, rare—insights. If you believe Darwin, nature is but a gigantic system of trial and error, of craftsmanship. Try this species. Oops. That didn't survive. Mutate. Try another.

Medicine, to take another example, is a centuries-old craft. Doctors treat patients by trial and error. A patient complains of headache. The doctor suggests Tylenol. It doesn't work. She suggests Fiorinal. It doesn't work. They do a CAT scan. It shows no abnormality. The patient is referred to a psychiatrist. It turns out the patient is a Craftsman and at work she's surrounded by Technocrats. She quits her job and the headaches disappear. What doctor worth his or her salt would have *started* with exploratory neurosurgery? Open up someone's head just in case? Quick fixes. Medical science as a whole learns slowly, by trial and error, and occasionally leaps forward with a new insight. It's a painstaking process. We try not to hurry new drugs onto the market, for example, because they may do more harm than good. Remember Thalidomide.

In fact, all science is built largely on trial and error. A genius like Copernicus comes along with a new insight and the next years and even centuries are spent doing what Thomas Kuhn called *normal science.* Craftsmanship. Working out the bugs and the specific implications of the new insight. Until another genius comes along and breaks the gestalt, and everyone has to unlearn what they thought they already knew.

Teaching is a craft. I know, because I'm still learning it every day from my colleagues. I have graduated from apprentice to journeyman, but they are the master craftsmen. You cannot *tell* someone how to be a teacher. He or she has to do it under (yes, I said *under*) the patient, tolerant guidance of a master. I have to watch them, listen to them, trust them, follow them. Oh, to be sure, there are a few born teachers, Artists who just have an instinct for it. But they're rare. You can be the most brilliant academic in the world and an utter flop at teaching. If someone, say, a Technocrat, comes along to try to tell me that he or she has a new way to revolutionize teaching, I will listen with a very skeptical ear. And if some futurist tells me that the classroom is about to be replaced by palmtop computers, I shall weep bitter tears of dread. And I will resist. Violently. Human beings are actors and creators. We don't have to follow the trends. We can change them. We can put the lie to them all.

Entrepreneurship, too, is largely a craft. Owner-managers of small and medium-size businesses are, for the most part, Craftsmen. They have translated experience, a skill, a métier, into a business, usually by identifying some relatively small but important innovation in product or service. The ones who innovate on a grand scale, the really big success stories, are invariably Artists. And among the millions and millions of startups, these big stories are rare indeed. Some of them get killed off prematurely when a Technocrat is given too much power to put some order into the shop.

Managing organizations and managing people is a Craft. Not an Art and most certainly not a Science. Rejuvenating organizations may be an Art. Managing things, or thinglike aspects of organizations, may be a Science. But the Craft of managing people takes wisdom, patience, experience, authority, conviction, realism. And what does the future hold? The best predictor of tomorrow's weather is today's weather. If it's sunny today, the odds are better than 50-50 that it'll be sunny tomorrow. Tomorrow, most of us are still going to go into the office, into the school, into the lab, into the factory. While the future designs itself, we still have to work and to live. What we've been doing for the past few years is probably what we'll be doing for the next few years. And that's good, not bad. We need stability for healthy change. I know that sounds paradoxical and maybe it is. But let me give an example of what I mean.

Heroin addicts commit crimes to support their habit. When they are in the throes of the addiction, when they need a fix, they can't think straight. Heroin substitution programs, like methadone, are designed to give the addict some stability, some calm, so that he or she can make a rational decision to change. When you're strung out, living in chaos and crisis, you do not behave rationally. In our organizations, we need the honest, trustworthy, calm authority of the experienced Craftsman to be able to behave rationally, to be able to recognize and promote artistry when it comes along, to be able to recognize Technocrats and to have the legitimate authority to keep them in their place, to keep their technical virtuosity and brilliance subservient to larger social goals. Former French prime minister Georges Clemenceau said that war was a much too serious thing to be left to the military. Well, managing is a much too serious thing to be left to self-styled professional managers.

So the Craftsman looks at the new tools, for that is what they are, coming out of the management literature and says, not no, but maybe. Reengineering? Flat structures? Maybe, I'll have my best Technocrat take a look at it. Alliances? Maybe, I'll give it some thought. The information revolution, you say. Maybe, but, writes T. S. Eliot again, "*where is the wisdom we have lost in knowledge, where is the knowledge we have lost in information?*"

Remember our Craftsmen Jeb, Robert, and Rodney? They said things like, "People are the organization's biggest asset," "You have to invest in the people and the profit follows," "You can't change everything overnight. It takes time to weld a group of people together." They say most of the things we want to hear. But they said maybe to the Technocrats' brilliant schemes and got fired for it. And the organization went belly-up because it changed its structures and its personnel too much—not too little. Not because it went too slowly, but because it went too fast. It lurched from one piece of shiny new rhetoric to the next.

The Craftsmen are the preservers and the mentors. Since when have realism and conviction become outmoded, old-fashioned? They're the ones who will give our young people, most of whom are budding Craftsmen, a chance to get started. They're the ones who will accept them into the organization even if they are not brilliant. They won't ask them if they are charismatic. They don't expect them to be. They will ask them what they know and whether they have the desire to learn how to do. They're the ones who will mentor, empower, and teach and train and invest in them. The Artists haven't got the time. And the Technocrats haven't got the inclination. People are what count. The wrong person in the wrong slot can turn a good idea into a disaster.

Conclusion

It was the best of times, it was the worst of times, it was the age of wisdom, it was the age of foolishness, it was the epoch of belief, it was the epoch of incredulity, it was the season of Light, it was the season of Darkness, it was the spring of hope, it was the winter of despair, we had everything before us, we had nothing before us, we were all going direct to Heaven, we were all going direct the other way. . . .

Charles Dickens, *A Tale of Two Cities*

Those words were written in 1859. Well, *Plus ça change, plus c'est la même chose*—the more things change, the more they stay the same. It still is, as it is in every age, the best of times and the worst of times. Such is the human dilemma. To live with doubt and ambiguity and uncertainty and to forge ahead anyway. As Malcolm Muggeridge said of his religion, faith is not the absence of doubt but its acceptance.

Sure, times are tough and confusing. The dollar is battered. Big, once-invincible corporate giants and governments are having to live in a world that smaller businesses and institutions, and the vast majority of citizens, have always inhabited. A world without guarantees and a world that demands a sharp pencil, vigilance, and constant innovation. Their technocratic illusions have caught up with them; things don't stay the same for very long. So, the wrenching downsizing, restructuring, is going on. Some of these giants will need Artists at the top to find a new vision. Some will need Craftsmen to put an end to their delusions of grandeur and to get back to basics. Those whose backs are truly to the wall may even need, temporarily, a hardnosed, brilliant Technocrat at the helm to chop and hack their way to survival.

And while all this is going on, people are being laid off—but not at the Microsofts. The new economy is designing itself, and, as always, the new jobs are coming from the new players. Some of them will be Artists, some Craftsmen. None will be Technocrats (or, if they are, they won't survive). In that respect, North America, with its high rate of new venture formation, is the favorite for those with betting instincts. I don't think we should throw in the towel.

When I decided to take this trip, this search, this research, it was because I was angry and frustrated. I saw the Technocrats. I saw what they were doing and I wanted to prove that their recipes, their smug self-satisfaction, their pious phrases, and their sermonettes were dangerous. I wanted to show what I had often seen, that artistry existed in management. That dreams sometimes come true. Like most people, I, too, saw the world in black and white—leaders versus managers, if you will. I didn't know, I couldn't know, that I would find the Craftsmen and discover that they had been made invisible by this polarization of our thinking. But isn't that what research is supposed to be about? Finding things. Looking into the past to shed light on the present.

But not to forecast it into the future. I don't know if, as the new millennium opens, we are at the dawning of a new age. If we are, I don't know if it will be a renaissance of craftsmanship and of Craftsmen—of skill, *métier,* realism, wisdom, conviction, and legitimate authority. But I hope so, because Craftsmen will give Artists room to breathe and time to create, and they will keep Technocrats in their legitimate place. I do know that we have a say in the matter. The future is not in a futurist's trend-line. The future is us.

Appendix A

What Kind of Leader Are You?

This book was, as you know, based on a doctoral thesis. Juries that approve doctoral work are interested in one thing and one thing only: methods. They are not particularly interested in the content of the results or in the conclusions of the researcher. What they judge is: Do the methods underlying the research justify the interpretation of the results? My research into character was built on 8 years of direct observation, 60 in-depth interviews, 30 years of internal and public documents, and 2 psychological tests—the Minnesota Multiphasic Personality Inventory (MMPI) and the modified Adjective Check List (ACL), which I call the Pitcher Adjective Test (PAT). While it is impossible for you to duplicate all these methods, the PAT will give you a guide as to what kind of leader you are, or what kind of person you're considering promoting, or the complexion of the whole management team (are the right types in the right jobs?). But it must be administered honestly.

Here's how to proceed. Give the PAT, as I did, to 10 colleagues. Make sure that they're not all your friends. As I was administering the test in interviews, I made sure that I had three Artists, three Technocrats, and four Craftsmen (for good judgment) analyze each CEO in order to minimize any bias. Even if you don't know for sure, you'll have a hunch about who's who. In any case, choose very different kinds of people. In addition, I tried to ensure a representation of peers, superiors, and subordinates. Covering peers and subordinates is most important to judge Technocrats because, on the whole, Technocrats are "nice" to superiors and very arrogant with subordinates. (In fact, one good way to identify a Technocrat is to find out how many secretaries or assistants he or she has had in the last five years. Usually, it's a revolving door. If I were hiring someone, I'd ask for letters of reference from his or her last two secretaries.) Make sure the process is confidential so that people can be honest. Ask some outsider, someone whose integrity is beyond reproach, to receive and compile the results and guarantee their anonymity.

What to Do with the Results

It depends. It depends on what you're trying to achieve and what position you occupy. If you want just to get an idea of how others see you, it can stop there. Or maybe you want to reorient your career, and you'd like to know what your strengths are. Or maybe you doubt the accuracy of your own judgment with respect to someone you want to hire or promote. Or maybe you're an entrepreneur building a management team and you want to know if there are significant talent or skills gaps in your organization. Or maybe you're a CEO considering whether you've got the right skills in the right place; for example, what kind of person do you need as a Vice President of Sales, or Production, or Marketing? Or maybe you're a board member who has to judge a new successor for the top slot. It depends.

The point is, asking someone to do what it is not in his or her nature to do is worse than useless. It's like trying to push a boulder up a mountain, or telling a child to grow up. You may order a plodding technocrat to come up with new ideas or to create a learning organization, or to encourage participation, or, or, or, but he or she can't. Similarly, you may order an pure Artist to design a new management information system, but don't hold your breath until it gets done. Want a brand new idea? Get an Artist. Want loyalty in your organization? Get a Craftsman. Want to cut a thousand people from your staff? Get a Technocrat. (But make sure he or she has the power to recommend but not the power to decide. Otherwise, you risk losing the very people who can help to build the future.)

You're Looking for Patterns

This is not a science. What you're looking for is not, for example, 10 out of 10 on *stiff.* You're looking for a gestalt, a portrait, a combination where both the highs and the lows and close substitutes (like *imaginative* and *intuitive*) count. The following table provides you with guidelines of what to look for.

Type	Highs	Lows
Pure Artist	Imaginative, daring, intuitive, exciting, visionary, entrepreneurial, unpredictable, funny, inspiring, emotional	Conservative, controlled, well-balanced, stable, detail-oriented, meticulous, methodical, distant, stiff
Gentle Artist	Imaginative, inspiring, entrepreneurial, intuitive, funny, easy-going, people-oriented, warm, humane, wise, generous, open-minded	Conservative, controlled, meticulous, cerebral, distant, stiff, uncompromising
Authoritarian Artist	Unpredictable, volatile, daring, imaginative, entrepreneurial, intuitive, analytical, determined, brilliant	Reasonable, stable, steady, conventional, amiable, helpful, humane, warm, distant, stiff, methodical

Type	Highs	Lows
Pure Craftsman	Responsible, sensible, honest, reasonable, trustworthy, realistic, predictable, steady, knowledgeable	Uncompromising, distant, stiff, difficult, cerebral, brilliant, volatile, exciting, intuitive, daring, bold, visionary
Regimental Craftsman	Straightforward, honest, responsible, conventional, well-balanced, predictable, conservative, controlled, methodical, knowledgeable	Exciting, bold, daring, intuitive, brilliant, difficult, emotional, imaginative, funny
Creative Craftsman	Sensible, trustworthy, knowledgeable, honest, wise, amiable, helpful, open-minded, people-oriented, generous, easy-going, entrepreneurial, inspiring	Cerebral, distant, stiff, uncompromising, intense, difficult, exciting, volatile, daring, unpredictable
Pure Technocrat	Uncompromising, hardheaded, cerebral, detail-oriented, determined, analytical, intense, no-nonsense	Honest, helpful, generous, open-minded, amiable, humane, warm, people-oriented, imaginative, intuitive, inspiring
Plodding Technocrat	Uncompromising, hardheaded, controlled, stiff, serious, conservative, methodical, detail-oriented, analytical, meticulous	Honest, wise, open-minded, generous, warm, inspiring, intuitive, exciting, daring, bold, brilliant

Type	Highs	Lows
Flashy Technocrat	Uncompromising, hardheaded, difficult, volatile, brilliant, cerebral, intense, determined, detail-oriented, analytical	Honest, realistic, sensible, trustworthy, reasonable, open-minded, generous, warm, imaginative, intuitive, inspiring, people-oriented

Pitcher Adjective Test (P.A.T.)

Subject: ________________ Observer: ________________

From among the following list of words, please check those that most apply to the person concerned. You may check as many words as you see fit.

1	[]	volatile	18	[]	conservative
2	[]	humane	19	[]	intuitive
3	[]	stable	20	[]	honest
4	[]	unpredictable	21	[]	distant
5	[]	well-balanced	22	[]	insightful
6	[]	controlled	23	[]	straightforward
7	[]	funny	24	[]	cerebral
8	[]	people-oriented	25	[]	brilliant
9	[]	serious	26	[]	hardworking
10	[]	imaginative	27	[]	sensible
11	[]	thoughtful	28	[]	difficult
12	[]	predictable	29	[]	wise
13	[]	daring	30	[]	responsible
14	[]	helpful	31	[]	exciting
15	[]	conventional	32	[]	warm
16	[]	entrepreneurial	33	[]	uncompromising
17	[]	amiable	34	[]	emotional

35 [] stiff
36 [] intense
37 [] trustworthy
38 [] meticulous
40 [] dedicated
41 [] detail-oriented
42 [] generous
43 [] punctual
44 [] visionary
45 [] easygoing
46 [] realistic
47 [] open-minded
48 [] analytical
49 [] entrepreneurial
50 [] knowledgeable
51 [] determined
52 [] steady
53 [] methodical
54 [] polite
55 [] reasonable
56 [] inspiring
57 [] fastidious
58 [] hardheaded
59 [] no-nonsense
60 [] other ________

Appendix B

Why Art?

In *The Courage to Be,* theologian Paul Tillich wrote:

> They [Kierkegaard, Nietzsche, Bergson] realized that a process was going on in which people were transformed into things, into pieces of reality which pure science can calculate and technical science can control. The idealistic wing of bourgeois thinking made of the person a vessel in which universals find a more or less adequate place. The naturalistic wing of bourgeois thinking made of the person an empty field into which sense impressions enter and prevail according to the degree of their intensity. In both cases the individual self is an empty space and the bearer of something which is not himself. . . . Idealism and naturalism are alike in their attitude to the existing person; both of them eliminate his infinite significance and make him a space through which something else passes.

I sensed that what Tillich said applied to management as well, but, outmanned and outgunned, in the beginning of my research I could do nothing but beat a strategic retreat. The first rule of combat is *Know thine enemy.* Well, that takes time. Fortunately, I had it. I set out to discover from whence came management's exceedingly limited vision of people. Of course, that meant reading "the greats." In management, they are Fayol, Taylor, Barnard, and Selznik; going forward in time, they are "the moderns," Harvard's Ken Andrews and Michael Porter, McGill's Henry Mintzberg, and USC's Warren Bennis. Peter Drucker and Herbert Simon straddle the two eras.

I got stuck at Herbert Simon. Little known in the outside world, Simon is a giant in management. You can feel his influence everywhere. *Bounded rationality* is his flag, and his flagship led a whole fleet of cognitive psychologists into management's quiet harbor. In 1960 he told us in his book *The New Science of Management Decision* that, unlike the management of yore, modern management would rely on training more than on selection, on information systems more than on intuition, on planning more than on judgment.[1] Below is an abridged version of a little table of his.

Notice the pejorative phrasing, juxtaposing *traditional* (bad) against *modern* (good); *heuristic computer programs* are modern, *judgment* and *intuition* old-fashioned. Spiritual descendant of Frederick Winslow Taylor, Simon took Taylor's principles of *scientific management* from the shop floor to the senior reaches of the organization. Michael Porter was later to take them right into the boardroom with his scientific management of strategy and the *one best way.* More on that later.

	Decision-making techniques	
Types of decisions	**Traditional**	**Modern**
Nonprogrammed: One-shot, ill-structured, novel policy decisions Handled by general problem-solving processes	1. Judgment, intuition and creativity 2. Rules of thumb 3. Selection and training of executives	Heuristic problem-solving techniques applied to: a. Training human decision makers b. Constructing heuristic computer programs

For the moment, I was stalled. I approached Simon with kid gloves. For as a child of the radical 60s and early 70s, I had learned a hard lesson. Those of us infatuated with Marx had adopted first Keynes and then John Kenneth Galbraith as our economic gurus. Milton Friedman was our diabolical archenemy. Then, one day in the late 70s, I went to listen to Friedman. This man was also a giant. The breadth and depth of his scholarship, the power of his analysis, were breathtaking. He made his left-wing critics and questioners in the audience look like intellectual pygmies. It was a very sobering experience. I realized Simon was a Goliath and I had no intention of tackling him with a slingshot. I wasn't sure that God was on my side.

All I knew was that Simon was wrong. Wrong about what? Everything. Wrong how? I don't know. You know how you can have a nagging, uncomfortable feeling about something but it's impossible to explain or define? Like the fairy tale where the "true" princess could only be discovered by putting a pea under a stack of mattresses and finding out whether or not she could sleep. Well, without comparing myself to a true princess, I couldn't sleep. So I went looking for causes. I started my search in Simon's bibliography: tell me who a person reads and I'll tell you what they think. Herbert Simon's mentors were, and are, cognitive psychologists. Maybe by reading cognitive psychology, I would discover the nature of the pea.

I turned for guidance to Howard Gardner's *The Mind's New Science,*[2] the somewhat pretentious title of a survey of cognitive science's accomplishments, and basically discovered that here people were, more or less (usually less), adequate *information processors.* They were a brain, a mind-machine without a heart, a Lockean "empty field into which sense impressions enter and prevail according to the degree of their intensity and. . . ." So, if a person is a mind-machine, it stands to reason that the machine can be "fixed" or perfected, and therefore training becomes a credible substitute for selection. Perfectly logical, you say. Well, yes, if the premise were true. But it isn't. People are not like machines; they have a heart and emotions, and that heart influences what they see and think. Cognitive psychology is an assembly kit that comes batteries not included: no morality, no emotion, no will to make the machine run. It freely admits that it had to set aside "the influence of affective factors or emotions, the contribution of historical and cultural factors, and the role of background context in which particular actions and thoughts occur."[3] Okay, so to proceed with any research we have to make some simplifying assumptions, but—come on—affect, history, volition, judgment, context, culture? What's left worth knowing?

In any event, all this stuff had been imported directly and unadulterated into management research. Human being as information processor is the subtle and sometimes not-so-subtle intellectual underpinning. Thus in strategy, for example, you don't need judgment or intuition if you have adequate information and you process it adequately. This notion is best exemplified by Michael Porter's enormously successful books *Competitive Advantage* and *Competitive Strategy.*[4] Here we find an exhaustive, and exhausting, recipe for collecting all the information managers should accumulate in order to deduce a strategy. If the right information is collected and the analysis is done right, anyone can sit in the executive suite no matter what their character or temperament. It's a useful recipe for novices, for students, and a wonderful and useful pedagogical device for professors of management. It's also a great tool for management consultants hired to develop a strategy for an organization operating in a sector of which they have no firsthand knowledge. And it's essential for peripatetic managers, who flit from one sector to another, for it gives them a way to acquire some superficial understanding of each new sector. But I have never met a seasoned manager who needed it, used it, or valued it. This became another clue as to what may be going on in the "science" of management.

Anyway, *The Mind's New Science* turned out to be a dud, and I'd taken a year-long detour down a blind alley. Nothing left but to retrace my steps and take Frost's road less traveled. If cognitive psychology was an assembly kit that came batteries not included (emotion and volition), I needed to find out more about affect, or feeling, and the role that it plays in human mentation, judgment, decision making. Some evidence emerged from experimental psychology.[5] One researcher has shown that emotion affects all kinds of judgments, decision making, strategy, and willingness to take risk:

> In combination with other data showing effects of positive affect on word-association, memory and creative problem-solving, these data support our hypothesis that positive affect is associated with cognitive reorganization such that more relations among concepts or ideas are seen than is the case under neutral-affect conditions.[6]

Here, we're a long way from the information-processing model. And yet, isn't managing all about "creative problem-solving"? Isn't "cognitive reorganization" exactly what we need, faced with the manifest failure of old solutions to competitive, and social, challenges? What literature could I explore that addresses this dicey, delicate, intricate relation between feeling and thought?

If modern cognitive psychology has largely conceived of the human as machine, depth psychology (or dynamic psychology) has been from the beginning more interested in the ghost that inhabits the machine. From the beginning, it was interested in emotion, affect, the batteries, what makes the system move. Psychopaths who murder little children do not lack information—they know they are breaking the law and all of our moral codes—they lack conscience. Someone suffering from paranoid delusions is not just in need of more information. And it's something of an understatement to say they have information-processing difficulties.

I decided to see how one of the best depth psychologists treats the relation between affect and cognition, or emotion and thought. David Shapiro describes the cognitive fixity and rigidity of the obsessive-compulsive personality and its relation to underlying affect:

> An obsessive compulsive patient—a sober, technically-minded and active man—was usually conspicuously lacking in enthusiasm or excitement in circumstances that might seem to warrant them. On one occasion, as he talked about a certain prospect of his, namely, the good chance of an important success in his work, his sober expression was momentarily interrupted by a smile. After a few more minutes of talking, during which he maintained his soberness only with difficulty, he began quite hesitantly to speak of certain hopes he had only alluded to earlier. Then he broke into a grin. Almost immediately, however, he regained his usual, somewhat worried expression. As he did this, he said, "Of course, the outcome is by no means certain," and he said this in a tone that, if anything, would suggest that the outcome was almost certain to be a failure. After ticking off several of the possibilities for a hitch, he finally seemed to be himself again, so to speak.[7]

It's as though a certain affect, or feeling—in this case joy or enthusiasm—were struggling to the surface against an implacable foe. The affect is unacceptable and leads to a series of thoughts that will dampen it, choke it off. Shapiro argues that the feeling of joy is associated in this particular man's mind with exaggeration and loss of control, an "unrealistic" optimism that could bring massive disappointment later. He has learned to control these dangerous feelings with thoughts. His thoughts make the feelings go away. From the outside—the vantage point of cognitivists—his information processing is impoverished, his thought processes illogical. But to him, they are lifesaving. He clings to them tenaciously. He thinks his logic impeccable and his prudence admirable. Often, he will believe that those who follow their impulses and enthusiasms are dangerous fools.

Most of psychoanalysis and psychiatry is devoted to discovering how this man and others like him got this way. The two professions are concerned with etiology and treatment. They're concerned with biological and genetic influences, maturational processes and child-rearing practices, all of which are fascinating but not for our purposes here. Further, because they grew within the bosom of the medical model, they are concerned largely with pathology and have precious little to say about creative, positive human productions. And leadership, whether you call it "charismatic," "transformational," or "visionary,"[8] is one of those creative human productions. Thus, while I would carry with me some psychoanalytic baggage—its work on the analysis of character—I was still seeking a terrain that put the Humpty Dumpty of emotion, thinking, and action back together again, but resulted in greatness, not in pathology. I turned to philosophy. Away from Locke and Hume, back to Plato, forward to Kant and Kierkegaard. From Kant to Croce, Santayana and Cassirer, Ruskin, Collingwood and Reid, Merleau-Ponty, Langer, and Sparshott. In short, I turned to aesthetics, philosophy of art.

Ruskin wrote that fine art is the domain in which the hand, the head, and the heart go together, humanity's "finest hour." No one, at least no one in their right mind, would try to pretend that an artist's emotions get in the way of his or her thoughts and actions. In aesthetics the whole person—willing, feeling, thinking, acting—is restored to us.

The Nature of Art

The work of aesthetics has been largely classificatory. The art historian finds or creates periods and labels them classical or romantic. Styles are identified: mannerist, impressionist, cubist. There is explication of symbols as the critic seeks to sort the artistic wheat from the mannerist chaff. There is much discussion of technique, the origins and use of devices like perspective. But philosophers, aestheticians, have struggled for millennia to define art, and the intensity of the debate has not cooled in our day.

Art as Revelation

If there has been controversy around what constitutes art, there is much more consensus around art's achievements and, indeed, its purpose. British aesthetician Sir Herbert Read insists that art has always revealed to us new insights:

> Art, in so far as it has retained its function and not become a matter for pastry cooks, has throughout history always been such a mode of revelation, of establishment, of naming.[9]

The artist, he argues, is an explorer for humanity and the "explorer's fumbling progress is a much finer achievement than the well-briefed traveler's journey."[10] His task is to *épater la bourgeoisie,* to break through the limitations of previously codified knowledge, to lead humanity into the future. Art, believed British painter John Constable, ought to be thought of as a branch of science; it investigates, it reasons. It creates symbols that facilitate thought:

> . . . a symbol is used to articulate ideas of something we wish to think about and until we have a fairly adequate symbolism we cannot think about them.[11]

Language provides that symbol system for discursive logic, but Langer tells us that there are whole domains of experience that philosophers deem *ineffable,* or too great for words. Art provides the medium, the language, for the expression of these ineffable thoughts. This may help us to better understand Robert Motherwell, who says, "All my life I've been working on the work—every canvas a sentence or paragraph of it." Or Isadora Duncan, who cried, "If I could say it, I wouldn't have to dance it."[12]

Gombrich, eminent art historian, provides a metaphor:

> The history of art, as we have interpreted it so far, may be described as the forging of master keys for opening the mysterious locks of our

> senses to which only nature originally held the key. They are complex locks . . . [and] like the burglar who tries to break a safe, the artist has no direct access to the inner mechanism. He can only feel his way with sensitive fingers, probing and adjusting his hook or wire when something gives way. Of course, once the door springs open, once the key is shaped, it is easy to repeat the performance. The next person needs no special insight.[13]

The next person needs no special insight; he or she may perfect or adapt the insight, as a craftsman does. Or mimic the insight, as a technician or a virtuoso does. In fact, great artists are always followed by a host of imitators—mannerists. Haven't we all seen pseudo-Monets at our local fairs? And pseudoleaders in our organizations?

The true artist's lot, however, is normally not an easy one. The great ones often burn out early, like Gauguin or Rimbaud. Each attempt to forge the key, to unlock the treasure, may be met with an infinite sense of disappointment, depression, even despair. Thus does a Giacometti destroy his sculptures almost immediately after their birth.[14] They fail him. They do not live up to their promise.

Artistic Perception

Art, writes Sparshott, occupies the ground between "chaos and cliché," and

> artistic activity begins when man grapples with the twisted mass of the visible which presses in upon him and gives it creative form. . . . The true artist cannot acquiesce in this phenomenal chaos.[15]

When we survey today's economic and social landscape, doesn't it seem like "a twisted mass"? Many aestheticians have sought the explanation for art in the perception of artists. For some reason, the artist is viewed as more susceptible to the world, more sensitive to its subtleties, more prone to see chaos where others see order. Some have attributed this partly to heredity, pointing to studies that show artistic activity runs in families.[16] Others have argued that the artist's perception is simply clearer, uncorrupted by concepts of what "should" be seen. Another aesthetician, Roger Fry, concludes:

> Art, then, is, if I am right, the chief organ of the imaginative life . . . and, as we have seen, the imaginative life is distinguished by the greater clearness of its perception, and the greater purity or freedom of its emotion.[17]

Perception is not, of course, simply that which strikes our sense organs. We bring structures to perception, and the explanation of art concerns itself as much with the artist's inner life as with his or her apparent reactions to the visible world. Here, as Fry notes above, the imagination is king. The artist is someone more attuned to the rhythms of the inner world—imagination, fantasy, dreams. The poet Spender writes that he is "aware of a rhythm, a dance, a fury, which is yet empty of words."[18] Coleridge apparently crafted his dreamlike *Kubla Khan* in a reverie.

The dreamlike quality of art is nowhere more evident than in modern surrealism. Whether it be stream-of-consciousness literature evident in Proust or James Joyce, or in the paintings of Marc Chagall (see the following reproduction), these works seem to capture not only the fleetingness of dream reality but also the strangeness of its forms and images. Time, space, and concept lose their imperious control over shape and substance. Our normal, everyday, conventional ways of seeing things are broken down.

Bennis tells us that one of the characteristics of transformational leaders is that they are adept at "unlearning." In general, great art always seems to involve both destruction and creation—destruction of old forms, old gestalts, in the service of the new, or, as Barron puts it, "the structure of the world must be broken then

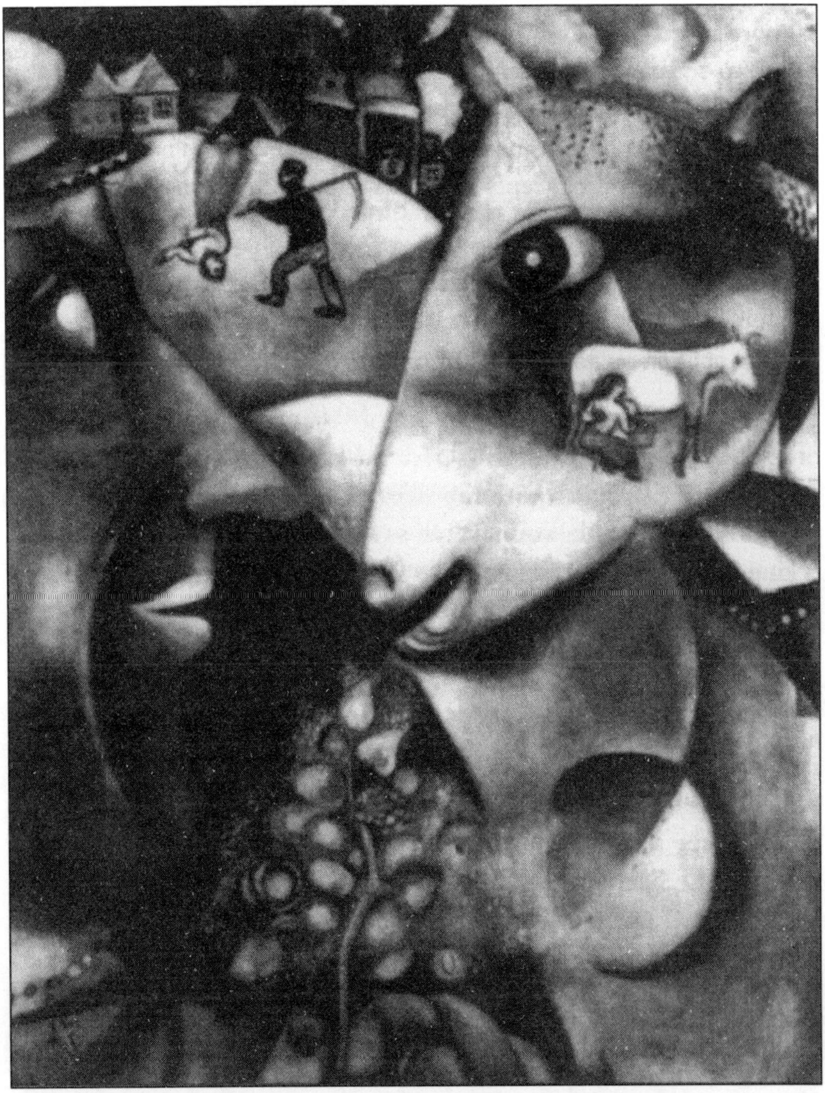

Figure B.1 Marc Chagall, *I and the Village*, 1911. The Museum of Modern Art, New York.

transcended."[19] Some artists seem to have understood this basic dialectic; listen to Picasso and Schiller:

> Thus, no less gifted an artist than Picasso confesses that, "a painting moves forward to completion by a series of destructions." And, the poet Schiller wrote, "What would live in song immortally / must in life first perish."[20]

To suggest that the everyday structure of the world must first be broken and the world of dreamlike images given precedence is not to suggest that art is but a dream or fantasy, a pure product of the unconscious. Consciousness controls the production of the creative product, but it seems to work on a fragment, a hint, a foreshadowing, not directly given to it. This is why the act of creation is anything but the tidy transcription of some prior mental image. It takes work, often agony. Spender says of his own poetic inspiration that there is "but the dim cloud of an idea which I feel must be condensed into a shower of words."[21] Ducasse stresses this open-endedness of art and the fragmentary nature of the inspiration when he writes that "the feeling that the work of art comes to embody may be there only as a germ ab initio."[22] More clearly still, French philosopher Merleau-Ponty suggests that because of the fragmentary nature of vision and its inarticulate nature, we never know in advance whether an artistic vision will be "useful" or not. We have to see the finished product:

> There is nothing but a vague fever before the act of artistic expression and only the work itself, completed and understood, is proof that there was *something* rather than *nothing* to be said.[23]

We saw that artistic leaders also have trouble explaining their visions and justifying their actions.

If the artistic enterprise is, then, an attempt to build a new intellectual home for humanity, a new order more suitable to our growing capacities, it must break with the past and shape new forms. To do so, it demands trust and time, two things that today are in short supply.

Finally I was getting somewhere. Vision is the currency of great art and of great leadership. Artists and leaders, and great scientists for that matter, are compelled to break with conventional wisdom; as Bronowski says, we expect them to "fly in the face of what is established."[24] But I still needed to know more about how the artist accomplishes his or her task. I needed to know more about the artist's character. Plato said: "The person who comes to the doors of poetry without madness from the Muses, convinced that technique will have made him an adequate poet, gets nowhere; the poetry of the self-controlled is annihilated by that of the crazed."[25] I already knew that mastering the techniques of modern management didn't make great managers, but did this mean that great artists and visionary leaders were all nuts?

Well, no, not exactly, but they are abnormal in the same sense that someone seven feet tall is abnormal. It is in the emotional domain that they distinguish themselves from equally intelligent but less creative people. They are nonconformist, like complexity and disorder more than others, and score much higher on tests of intuition. They are moody. They have a marked tendency to depression on the one hand or to hyperthymia (elevated mood) on the other, or to a cyclic combination of the two—cyclothymia. Da Vinci's depressions were legendary; van Gogh's, lethal. Lord Brain, a British aesthetician, tells us about Charles Dickens, whose moods alternated regularly between elation and despair:

> I have suggested before that Charles Dickens was a cyclothyme. His prevailing mood was one of elation which gave him his enormous gusto and energy which, apart from his prolific inventiveness, was always overflowing into long walks, parties and amateur theatricals. His son, Sir Henry Dickens, speaks of his father's "heavy moods of deep depression, of intense nervous irritability, when he was silent and oppressed." Dickens himself [wrote] "I am sick, giddy and capri-

> ciously despondent. . . . I have bad nights, am full of disquietude and anxiety . . . my wretchedness just now is inconceivable."[26]

I expected to discover—indeed some of the leaders I was studying had already said as much—that some great leaders were subject to depression and discouragement. After all, defying conventional wisdom is not an easy job. But, like Dickens, they are also characterized by prolonged periods of immense energy and boundless optimism. The flip side of the cyclothymic character, this provides the drive, the enthusiasm, the force to impose the new vision.

Most people are wedded to the conscious gestalts, or the *cognitive schemas,* of everyday life. In general this is good. It keeps us on track. But as Jung reminds us, a track can become a rut without our artists:

> The definiteness and directedness of the conscious mind are extremely important acquisitions which humanity has bought at a very heavy sacrifice, and which, in turn, have rendered humanity the highest service. Without them science, technology, and civilization would be impossible, for they all presuppose the reliable continuity and directedness of the conscious process. . . . We may say that social worthlessness increases to the degree that these qualities are impaired by the unconscious. Great artists and others distinguished by their creative gifts are, of course, exceptions to this rule.[27]

The artist has better access to the imagination, but so does the schizophrenic. Unlike the schizophrenic, however, the artist's conscious faculties are able to control and dominate the flood of imagery and associations originating in the unconscious.[28] Thus, he or she is able to break with conventional gestalts and ideas and forge new ones, whereas the compulsive personality, so afraid of letting go, has his or her consciousness dry up into a "banal, empty, arid rationality."[29]

All right. I won't trouble you with any more of this. By now you have enough to go on. The artistic character is quite distinct. I also had got some clues about its opposite, "banal, empty, arid rationality." And along the way, I had learned about craftsmanship.

The Artist Ideal Type

The object of my voyage into the domain of art was to discover the character of the artist and the nature of his or her enterprise on the assumption that it was not dissimilar to visionary leadership. In 1905 the philosopher George Santayana told us, "Man's progress has a poetic phase in which he imagines the world, then a scientific phase in which he sifts and tests what he has imagined."[30] We depend, as a civilization, on the artists' vision; it is they who force us, often reluctantly, to change our ways of seeing. Their visions stem not from some conscious desire to be rebellious, but from their character. They are peculiarly susceptible to the outer and inner world; they live, precariously, at their frontier. My search permitted me to construct the Artist Ideal Type, a set of descriptors that might represent, let us say, every artist in general and none in particular. Lord Brain's "administrative artist," I hazarded, would evidence a frustration with conventional wisdom, a willingness, even a need, to depart from it. This would stem from his or her character, a character that would have a hyperthymic or cyclothymic base and that would cause him or her to be subject to bouts of depression and periods of elation. In those periods of elation, he or she would have lots of ideas. To others, he or she will seem intuitive, imaginative, unpredictable, volatile, emotional; some may believe he or she lives in a dreamworld.

The Technocrat Ideal Type

To see the Artist clearly, we have need of a contrast. When technique is elevated to a first principle, its carrier becomes a Technocrat, "one for whom the technical sides of an issue take precedence over the social and human consequences."[31] If what predominantly characterizes the artistic character is the capacity to alternate between consciousness and the unconscious, to live on the frontier without becoming ill, then what characterizes the opposite is that he or she cannot do so and, worse, disdains those who can and do. Earlier, Rollo May referred to the compulsive character, those who are so fearful of unconscious contents that their consciousness dries up in "banal, empty, arid rationality." If the Artist is typified by a reaction against conventional wisdom, the compulsive character clings to it, conforms to it, deifies it. In trying to impede the flow of unconscious material, compulsives succeed in stultifying all affect, all emotion. In general, they exude emotional control, even rigidity. Reich called these people "living machines."[32] Machines don't get hunches; Technocrats will rarely be described as intuitive. What counts for them are the "facts," the "rules," the "right way to do things."

Playfulness is not on the agenda for Technocrats. Spontaneity is ruled out. They are stubborn and will brook no interference with their plans, which are, after all, based on the "facts." Their chief defense mechanism is intellectualization. Because they are most comfortable with detail, with facts, they often evince technical virtuosity and ingenuity, and this is the source of their power. In management, we need only think of the accomplishments of Frederick Winslow Taylor, his obsessional time-and-motion studies, to see both the virtues and the difficulties of this character pattern. His system was ingenious, but he missed the affective tone of the controversy it engendered. For him, it was simply based on the facts, which were, for him and by definition, emotionally neutral. Of course, facts have no status independent of time and of their observers; it used to be a fact that the world was flat. And a person, or a "society which believes in magic, witchcraft and oracles, will agree on a whole system of 'facts' which modern men regard as fictitious."[33]

But the Technocrats are not just compulsive, they are also narcissistic. Their way of doing things is the *only* way.

The Craftsman Ideal Type

If the Artist's visions are "bridges thrown toward an unseen shore,"[34] the Craftsman is the guy who builds the bridges. In an organizational context, Mintzberg has taught us about crafting strategy.[35] In keeping with the work of Lindblom and Quinn, Mintzberg sees strategy formation as a continuous, incremental process, one that necessarily adapts to the continuous demands of an uncertain marketplace. In general, craftsmanship is always *étapist.* It is always about the constant, slow accumulation of knowledge and skill. As noted earlier, the distinction between craft and art continues to provoke considerable debate within aesthetics. When does craft become art? Are they two distinct categories or do they exist on a continuum? These are just some of the questions that plague philosophers. Our interest is not this debate but rather the set of descriptors about craft the debate provides, allowing us to define the Craftsman Ideal Type.

If, with Santayana, we believe that human progress has a poetic phase, which corresponds to art, and a scientific phase, which here corresponds to technique, it also has a craft phase, where what is found with art comes into use and is transformed, perfected, refined, concretized, shaped, sculpted by experience. It could be argued, for instance, that the French Impressionist painters created a genre in which the Craftsmen, the Canadian Group of Seven, worked. They adapted the peculiar insights of the French to a uniquely Canadian context. Samuel Johnson, the great British satirist, was undoubtedly wrong when he said, "You cannot with all the talk in the world, enable a man to make a shoe."[36] Writing before the triumph of industrialism, he could not foresee that what was then a strong craft could, with time and ingenuity, be transformed into a set of techniques and that machines would one day make shoes. His underlying point about craft remains forceful, however. Craft is based on lore, a body of traditions.[37] These traditions pass from master to apprentice in an active, physical sense—not in a classroom or a book. You cannot "tell" a person how to make good pottery, a serviceable shoe, a watercolor image that does not run into the next, how to run an organization. These accomplishments require practice, long experience, Michael Polanyi's "tacit knowledge."

Craft is based on practice, and practice is not mere repetition. Practice, apprenticeship, teaches simultaneous doing and thinking. It is conceptual, as well as practical. "The procedures we follow tend to emerge not only from tradition but also from our momentary successes and failures in applying what tradition has given us."[38] Craft is the difference between *knowing* and *understanding;* the craft of singing, for example, is the "science of acoustics [knowing] joined to the experiential understanding of the body [understanding]."[39] Experience is essential to understanding.

Craftsmanship poses deep challenges in our modern world. A respect for the practical sides of any art demands submission to authority because the craft must be transmitted from master to apprentice. Polanyi insists:

> To learn by example is to submit to authority. You follow your master because you trust his manner of doing things even when you cannot analyze or account in detail for its effectiveness. . . . A society which wants to preserve a kind of personal knowledge must submit to tradition."[40]

Having submitted to learning by doing, we acquire a practical understanding of the art; it cannot be reduced to maxims. Polanyi continues:

> Maxims are rules, the correct application of which is part of the art which they govern. The true maxims of golf or poetry increase our

> insight into golf or poetry and may even give valuable guidance to golfers and poets; but, these maxims would instantly condemn themselves to absurdity if they tried to replace the golfer's skill or the poet's art. Maxims cannot be understood, still less applied, by anyone not already possessing a good practical knowledge of the art.

Most business schools require that MBA applicants have at least a modicum of experience before admission; this, because they sense that their maxims cannot be understood, still less applied, by anyone not already possessing a good practical knowledge of the art. On the other hand, how often have we been told a set of maxims about how organizations should be managed? And how often have these maxims been offered by people who have never set foot in our organizations? Further, as a society, we don't much like authority or tradition; so, we have a lot of problems with craftsmanship.

If the foregoing presents an adequate portrait of craft, what does it tell us about organizational Craftsmen? For one thing, they will be patient. To submit to authority implies an ability to set aside, temporarily, one's own preferred way of doing things. They will probably be conservative, in the sense of respecting tradition. They will not be quick to depart from the tricks of the trade. They will be frustrated by maxims detached from an understanding made possible only by experience. They will be responsible, sensible, and exhibit good judgment. They will be honest, loyal, and straightforward. They will be amiable and tolerant.

Well, this has been a rather long-winded attempt to put you into my frame of mind, to show you *what* I was looking for and *why*. My challenge became *how* to take these insights into management, how to study administrative genius, Artists in action, their fellow travelers, the Craftsmen, and their opposites, the Technocrats.

Previous leadership research has foundered largely on methodological reefs. First, researchers have confounded the position with the person who occupies it and concluded that all leaders are different. But not all people in positions of power are *leaders.* Some may have inherited the position. Some may have stolen it. Some may have been compromise candidates. Second, researchers have confounded public reputation with leadership. I know from bitter experience that many so-called visionary leaders with sterling public reputations do not merit the title: They've inherited the fruits of previous real visionaries' work. Finally, researchers have overestimated the value of interviews with "leaders." People dissimulate in interviews and, unless you are a psychiatrist or a psychoanalyst, it's very hard to detect. That's why to get a handle on leaders you have to interview them, to be sure, but you also have to interview their colleagues and subordinates—witnesses. And that takes a long time and multiple methods.

So, with these pitfalls in mind, here's what I expected to find: Artists, great leaders, who would be high on vision and capable of consciously controlling that vision; Technocrats, who would lack vision but would be very much oriented toward control; and Craftsmen, with, let us say, medium-term vision and sufficient control. I didn't expect to find any fools or lunatics. No fools get to the top of major organizations. Nor do many lunatics.

Multiple Methods

What I may call art, someone else may not. Someone I claim is an artist may be seen by others to be a fool or a lunatic. Somehow I

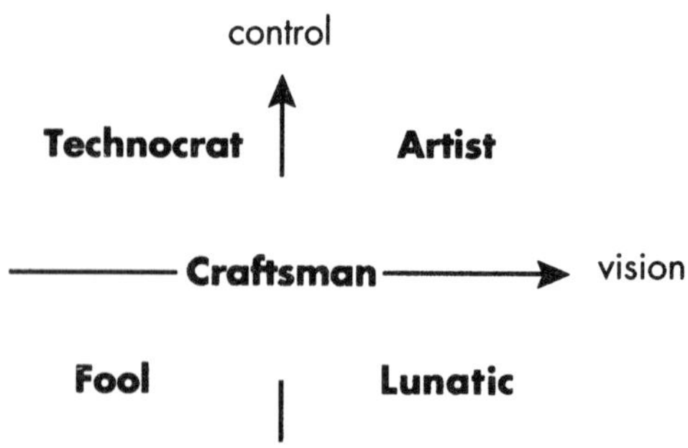

had to develop a set of research tools that would permit me to say certain things with some level of confidence, some level of inter-subjective validity. If you're still with me, here is what I did.

For many years, I watched inside a major international financial services organization with assets in the tens of billions of dollars and operating divisions on several continents.[41] As a witness at countless board and committee meetings, as well as special sessions, it was plain to me that there were three character types who were involved in the life of the organization at its highest level—the 15 CEOs of the group as a whole and of its major operating divisions. Each type spoke a different language, had different concerns, different priorities. One group wanted the organization to conquer the world, or at least major parts of it; another wanted to get better and better at doing what it already did; a third wanted to get the "systems" right and produce profit. Of course, both of the first two groups also wanted to produce profit, but their definitions or their notions about *how* to produce it differed enormously from the third's. How could I prove that there were indeed three distinct groups, that this was not just my fantasy? Three ways.

I indulged in the North American penchant for measurement. First I constructed something called an adjective checklist (ACL).[42] This list consisted of 60 everyday adjectives like *funny, serious, honest, controlled,* and so on. I expected Artists to be seen, among other things, as *intuitive,* Craftsmen as *realistic* and Technocrats as *meticulous.* In interviews with board members, peers and, in some cases, vice presidents, I then asked 10 observers to fill in these ACLs on each of 15 key executives. For example, James was the CEO of the group from 1965 to the early 1980s. I asked 10 of his closest colleagues—admirers and detractors—to respond to the lists with the instruction "When you think of James, what are the adjectives that spontaneously spring to mind?" The responses were collected and the result was 9,000 (60 adjectives × 10 observers × 15 executives = 9,000) bits of information. I then fed these data into a computer and got some fancy mathematical solutions—factor and correspondence analysis—that told me there were indeed three groups of people (see Appendix C).

Appearances can be misleading. This result, while encouraging, was based upon perceptions. What if the perceptions were wrong in some sense? What if the *real* person was different than he or she seemed to be? To corroborate the adjective story, I needed to know something about the inner world of those 15 people. Would the inner conform to the outer? So, I administered the Minnesota Multiphasic Personality Inventory (MMPI). This is a standard psychological test used every day in tens of thousands of clinical diagnostic settings around the world. It is not a quick and dirty cute little personality quiz. It is a serious instrument based on 566 questions, and takes several hours to complete. The results of this test were machine scored to guarantee objectivity and interpreted with the assistance of a clinical psychologist from Toronto. The personality profiles that emerged confirmed that I was dealing with a stable and consistent phenomenon. Once again, the three groups emerged.

For the rest, I studied internal and public documents for the 30-year period 1960 to 1990, with particular concentration on the years 1975 to 1990. In addition to watching, listening and talking, I conducted 60 in-depth interviews with executives and board members—two rounds in 1986 and again in 1990. I asked them about the evolution of the group, its strategy, its structure, its culture, its team spirit.

Appendix C

Technical Material

The adjective checklist (ACL) was administered in the interview context. Observers—peers, subordinates, and board members—were asked to check as many adjectives as they thought applied to the 15 target subjects—all CEOs. They were instructed to respond quickly and spontaneously to the stimulus words, keeping in mind dominant traits of the person in question. For example, although most people may be characterized as punctual, it is not necessarily something that automatically springs to mind with certain individuals. The objective was to delineate core characteristics. All subjects cooperated fully. Administration was adjusted continuously throughout this phase of the research to ensure balance in representation and to guard against systematic bias. That is, as results were analyzed successively, each target executive was judged by an equal mix of Artists, Craftsmen, and Technocrats. Complete administration resulted in a database of 9,000 responses.

Analysis

When all the results were tallied, I was confronted with a matrix of 9,000 data points. Visual inspection of the data seemed to confirm, in general, the hypothesis of three distinct groups. However, especially with a database of this size, perceptual bias is a serious obstacle; recency, saliency, prior schema processing all potentially interfere with or distort the emerging portrait.[1] Anomalies can be overlooked, important secondary trends, obscured. The data were, therefore, subjected to more formal analytic techniques, namely, factor analysis and correspondence analysis.

Factor Analysis

Factor analysis is a technique in widespread use; Kerlinger calls it "the queen of analytic methods."[2] Used as a data-reduction technique to search for mysterious underlying uniformities or clusters, or as an hypothesis-testing device to confirm or infirm a priori notions, it has proved to be an invaluable tool. In the present case, it offers a way to circumvent the perceptual bias noted earlier. The conceptual and mathematical underpinnings of factor analysis are well explained elsewhere,[3] but in lay terms the technique consists of using correlation analysis to compute constructs or *factors* underlying several variables or tests.

In standard R-type factor analysis, we would reduce our 60 adjectives into several groups that are highly correlated internally but weakly correlated with each other. Our interest, however, lies more with finding resemblances, not between adjectives, but between people. That is, we wished to test the hypothesis that there was a determinate number of groups of individuals who could, within the constraints of the definitions cited earlier, be described as Artists, Craftsmen, and Technocrats. This necessitated performing Q-type factor analysis,[4] reversing rows and columns in the matrix, replacing variables with people. Overall and Klett speak to a controversial point with respect to Q-type analyses:

> . . . the partitioning of individuals does not appear at first blush to make much sense. Within the context of the linear factor model, however, it is quite reasonable to conceive of "person factors" as ideal types and the factor loadings as indices of relationship of individuals to the several ideal types (p. 203).

The factor analysis was accomplished using the statistical package for the social sciences (SPSS) factor program with principal components extraction and rotation to simple solution. A *simple* solution is one in which each variable, in our case, each person, is loaded mainly on one factor, and there are many negligible loadings in the factor matrix.[5] We present here, in summary form, the last steps. Initially we find the global results and their levels of *significance,* followed by the final *oblique* solution.

The first thing to note is that, with a generally accepted cutoff limit of 1 for acceptable eigenvalues, the procedure resulted in three factors that together accounted for 75.1 percent of the overall variance. Table C.1 presents the results.

It is clear that we had three strong factors with solid eigenvalues. It remained to rotate the data to simple solution. An oblique rotation presumes that the factors are somewhat correlated. As

Table C.1 Initial Statistics

Variable	Communality	Factor	Eigen-value	Variance (%)	Cumulative Variance (%)
Jim	1	1	5.24322	35.0	35.0
Cobb	1	2	3.56595	23.8	58.7
Mike	1	3	2.46318	16.4	75.1
Ross	1	4	.91179	6.1	81.2
Pete	1	5	.58398	3.9	85.1
Cam	1	6	.55231	3.7	88.8
Judd	1	7	.35192	2.3	91.1
Brien	1	8	.31449	2.1	93.2
David	1	9	.25667	1.7	95.0
Rodney	1	10	.15990	1.1	96.0
Robert	1	11	.14663	1.0	97.0
George	1	12	.13759	.9	97.9
Rowan	1	13	.12562	.8	98.8
Bill	1	14	.10685	.7	99.5
Jeb	1	15	.07989	.5	100.0

between orthogonal and oblique rotations, there is no generally-accepted, "correct," procedure, it being largely a matter of taste.[6]

The Final Solution

As predicted, the pure craftsmen, Robert, Jeb, George, and Rodney, loaded highly—respectively, .94, .92, .90, and .86—and solely, on Factor 1, which we dubbed the craft ideal type.

Cobb, Mike, and Jim loaded, respectively, .88, .87, and .78 on Factor 2, which we called the art ideal type. Jim shows a lesser but still interesting loading on the craft factor at .31, which confirms that the artist may contain significant craftlike qualities.

As expected, David, Ross, and Peter load highly on Factor 3, the technocrat ideal type, measuring, respectively, .83, .68, and .67. Ross shows interesting loadings on the art factor at .32, indicating that a Technocrat may have artistic leanings or elements, and loads negatively on the craft factor at –.34. He is the only Technocrat to load negatively on craft, but he is not the only one to display anticraft characteristics. The two other Technocrats, loading .65 and .59, respectively, on Factor 3, are Cam and Judd. Their results are fascinating since they load almost as strongly on the art factor but in a negative direction. Their results show most strongly the polarity between the two ideal types.

Bill and Rowan were somewhat enigmatic during observation and interviews such that it was unclear whether they were predominantly Artists or Craftsmen. The results here confirm the earlier confusion, with Bill loading quite highly on craft at .84 but loading significantly, at .36, on Factor 2, art. Rowan, on the other hand, loads only on craft, but more weakly, at .52. This result was unsurprising in a sense, since it became clear during the interviews that there were radically different opinions of him. The results therefore tended to cancel out.

Finally, the anomalous (in the sense of unpredicted) result for Brien. He was predicted to fall firmly in the Technocrat ideal type but, in fact, loads more highly on the craft factor at .68 compared to .40. However, like the other Technocrats, Cam and Judd, he loads highly negatively on art at –.55. Table C.2 presents the results.

Discussion

These results were, globally, highly satisfying. Three strong factors, and not six or seven or two, emerged as hypothesized. The individuals who loaded on the factors were, with one exception, predicted. Anomalies are, of course, unsurprising. We are, after all, dealing with people and not with objects. The factors represent ideal types to which real individuals will be expected to conform only more or less. To understand the anomalies one is forced back to the raw data. When we compare, for example, the portrait

Table C.2 The Oblique Solution and the Ideal Types

Variable	Craftsman	Artist	Technocrat
Robert	.93262		
Jeb	.91573		
George	.89821		
Rodney	.86353		
Bill	.83884	.32296	
Brien	.68253	–.54930	.40238
Rowan	.52813		
Cobb		.88454	
Mike		.87064	
Jim	.31343	.77761	
David			.83254
Ross	–.33603	.32420	.68479
Peter			.67010
Cam		–.62962	.64522
Judd	.40150	–.58312	.58617

NOTE: Loadings under .3 have been suppressed for improved legibility.

of a pure craftsman with that of Brien, we discover that they are correlated at .6950 ($p < .01$). But, and this is more important, along what dimensions are they correlated or, more precisely, in what ways are they both alike and dissimilar? It turns out that both Brien and Judd share with Craftsmen their conventionalism, their emotional stability, but depart radically from them with respect to the qualities of warmth, humaneness, and generosity. This leads to a more general point about the factor analysis. As helpful as it is and as gratifying the results seem to be, it nonetheless left too many questions unanswered. Because we cannot, simultaneously, view the people, the ideal types, and the adjectives that served to create those types, we are forced into a still-intuitive and visual inspection of a large database. It is for this reason that we turned to a less well-known (in North America) European analytic technique called *correspondence analysis*. Its graphic outputs allowed us better to explain both the factors and the anomalies.

Correspondence Analysis

The conceptual and mathematical bases of correspondence analysis are explained, in detail, elsewhere.[7,8] In short, it uses a distance measure to allow the simultaneous representation of two data sets and is best applied to data consisting of contingency tables and binary coding.[9] Our data are precisely of this contingent nature; for example, "if you are Brien, how likely are you to be seen as visionary?" The output of correspondence analysis is also factors, but factors that have been created simultaneously by the weight of both the adjectives and the individuals. We are able to see the mathematical contribution of each individual and each adjective to the factor and the final graphic display places all the data before our eyes. We will proceed as we did with factor analysis, reporting first the overall statistics and moving to successively greater detail.

Initial Statistics

Like principal components analysis, correspondence analysis produces factors and associated eigenvalues. It does not, however, cut off the creation of factors according to the strength of those eigenvalues. In these initial statistics we see, therefore, all conceivable factors even though they contribute little explanatory power with respect to the overall variance in the data (see Table C.3). The principal thing to note is that, as in the factor analysis, the first three factors account for the same amount of variance, around 75 percent. The first two factors are overwhelmingly dominant, accounting for 42 and 27 percent, for a total of 69 percent.

As with principal components analysis, correspondence analysis requires knowledgeable interpretive effort to bear fruit. Unlike its North American counterpart, it does not use rotations to

Table C.3 Correspondence Analysis Initial Statistics

Factor	Eigenvalue	Variance (%)	Cumulative variance (%)
1	0.18360	41.759	41.759
2	0.12068	27.449	69.208
3	0.02854	6.490	75.699
4	0.02305	5.244	80.942
5	0.01448	3.293	84.235
6	0.01410	3.207	87.442
7	0.01189	2.704	90.147
8	0.01057	2.405	92.551
9	0.00752	1.711	94.262
10	0.00681	1.550	95.812
11	0.00567	1.289	97.101
12	0.00523	1.189	98.290
13	0.00415	0.945	99.235
14	0.00337	0.765	100.00

arrive at the best solution. The interpretive effort lies, instead, in inspection of the composition of the factors, the contributions of both rows (the adjectives) and columns (the people), to their creation, and inspection of the graphic representations of the factors, two-by-two, in two-dimensional space. The whole of that interpretive process will not be presented here. Here we present the final results.

What and Who Created Factor 1?

It is important to remember that with this analytic technique both positive and negative contributions count. That is, certain adjectives are mathematically attracted to each other and repelled by others. Similarly, certain people are mathematically attracted to each other and others are repelled. A factor in this case is a linear vector in space whose position is influenced by all other factors, individuals, and adjectives. We have seen from the initial statistics that Factor 1 is highly significant, accounting for more than 40 percent of overall variance in the data. Factor 1 is, let us say, composed of adjectives and people. In terms of mathematical weight or mass, it was created by, in order of importance, Cobb, Mike, Jim, and Ross and influenced negatively by, in order, Brien, Rodney, Judd, Robert, and Cam. It is clearly our Art vector or ideal type. With the positive contribution of our Artists from the factor analysis and the negative contribution of our technocratic ideal type, it shows even more clearly the polarity between the two styles. Interestingly, it also shows the negative contribution of two Craftsmen. The reason for this will become clearer shortly, but, for now, let us say that the Craftsman "repels" the Artist because, in some important respects, they are very different. Table C.4 presents the summary results.

As was true for people, so it is true for the adjectives. Some adjectives associate positively with the Art vector and others, negatively. With positive associations we find, in order, *bold, daring,*

Table C.4 The People and Factor 1

Positive		Negative	
Person	**Contribution**	**Person**	**Contribution**
Cobb	232	Brien	–106
Mike	157	Rodney	–88
Jim	101	Judd	–71
Ross	79	Robert	–69

NOTE: The software ADDAD uses a base of 1,000.

intuitive, exciting, volatile, unpredictable, and *entrepreneurial.* Conversely, *conventional, meticulous, controlled,* and *methodical* are strongly repelled. This begins to explain why the Craftsmen, Rodney and Robert, should have contributed negatively to the factor. A Craftsman is seldom described as bold and unpredictable. The Technocrat, on the other hand, is often described as conventional and, above all, meticulous. It is the richness of the simultaneous treatment in correspondence analysis that allows us to capture these nuances. The results are presented in Table C.5.

These results are, in themselves, fascinating, but they become more illuminating with each successive juxtaposition of other

Table C.5 The Adjectives and Factor 1

Positive		Negative	
Adjective	**Contribution**	**Adjective**	**Contribution**
Bold	71	Conventional	–39
Daring	55	Controlled	–34
Intuitive	54	Meticulous	–31
Exciting	51	Methodical	–30
Volatile	46		
Unpredictable	43		
Entrepreneurial	41		

results. Factor 2, as we have seen, contributes importantly to the explanation of overall variance. Its shape must therefore have much to tell us. We proceed to an interpretation of those results followed immediately by the juxtaposition of the two factors in space. Factor 3 will not be elucidated since it does not contribute materially to our understanding.

What and Who Created Factor 2?

Factor 2 contributes 27 percent to the explanation of overall variance. It is shaped, largely, by 10 individuals and 10 adjectives. In terms of positive contribution, and once again in order of importance, it is formed by Cam, Ross, David, Judd, and Peter—our Technocrats from the SPSS analysis. The vector is influenced negatively by Bill, Jeb, Robert, Jim, and George, three pure Craftsmen, one Craftsman with an artistic strain and one Artist with craftlike humanism. Here we see the polarity between the Technocrat and the Craftsman along, let us say, a temperament dimension. Table C.6 gives the highlights.

The individuals who influence the vector tell one kind of story but the adjectives embellish and enliven that story. In the case of Factor 2, the five strongly, positively associated adjectives are *difficult, hardheaded, distant, stiff,* and *uncompromising;* those

Table C.6 The People and Factor 2

Positive		Negative	
Person	**Contribution**	**Person**	**Contribution**
Cam	211	Bill	–98
Ross	140	Jeb	–84
David	100	Robert	–62
Judd	70	Jim	–54
Peter	63	George	–36

most strongly negatively associated are *humane, amiable, warm, helpful,* and *generous.* The vector appears to produce a sort of thermometer, a continuum, which ranges from warm to cold from the most pure Craftsman to the most pure Technocrat. Table C.7 shows the relative contributions.

It is worth emphasizing that these analytical procedures are producing constructs, ideal types, and that real people would seldom find themselves positioned neatly along one or another continuum. Constructs do, however, help to impose useful order on an otherwise incomprehensible world. Several constructs or ideal types taken together can help us to position and understand more clearly the real individuals who have made up this organization. Placing factors one and two together in two dimensions adds necessary and revealing complexity to the portrait we are drawing, and correspondence analysis, treating as it does all variables equally and simultaneously, permits us to do just that. Figure C.1 shows Factor 1, the art/technocracy vector, on the horizontal axis and Factor 2, technocracy/craft vector on the vertical axis. It displays both individuals and adjectives.

A word of caution about interpretation. In correspondence analysis, it is legitimate to compare clusters of adjectives and to compare, within clusters, their relative positions. Similarly, it is legitimate to compare individuals' positions within and between

Table C.7 The Adjectives and Factor 2

Positive		**Negative**	
Adjective	**Contribution**	**Adjective**	**Contribution**
Difficult	64	Humane	–41
Hardheaded	57	Amiable	–36
Distant	50	Warm	–31
Stiff	46	Helpful	–30
Uncompromising	38	Generous	–30

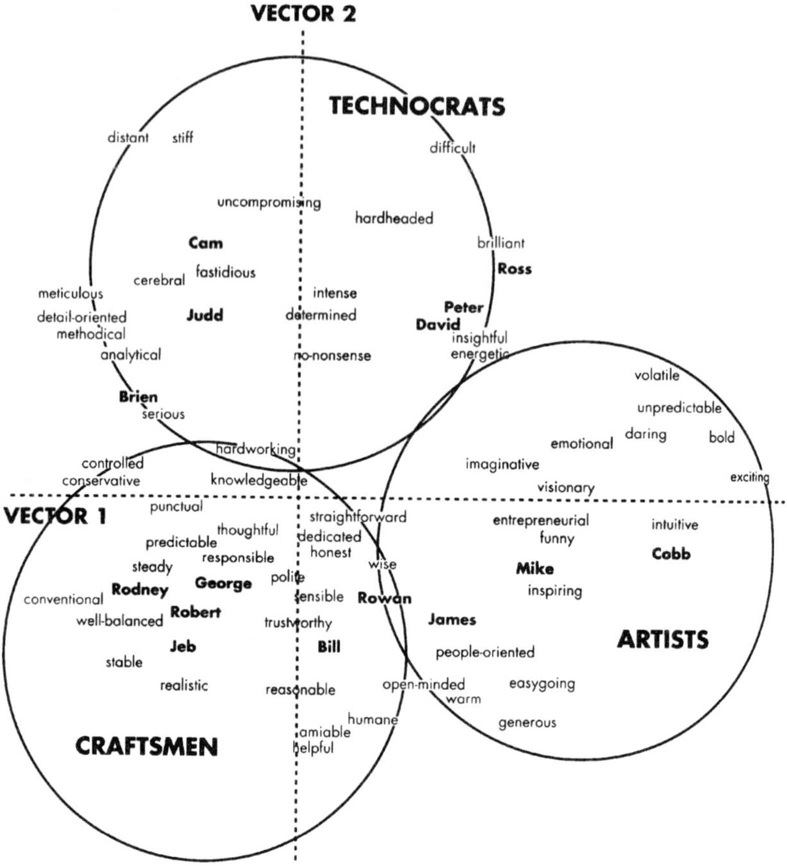

Figure C.1

clusters. And, it is legitimate to interpret individuals in relation to a group of adjectives. It is not, however, appropriate to speak to one individual and one adjective; that is, we cannot say "individual X is *stiff*" because the position of both the individual and the adjective is determined jointly by all the adjectives and all the individuals.

Notes

Introduction

1. Chester Barnard, *The Functions of the Executive* (Cambridge: Harvard University Press, 1938).
2. George Santayana, *Le Petit Robert* (Paris: Dictionnaire le Robert, 1990).
3. In Alain-Gérard Slama, *L'Angélisme exterminateur* (Paris: Bernard Grasset, 1993), 194.
4. Charles Taylor, *The Sources of the Self: The Making of the Modern Identity* (Cambridge: Harvard University Press, 1989), 512. Charles Taylor is a very "serious" ethical philosopher, but he uses metaphors that resonate, that touch our emotions as much as our intellect, because understanding begins in feeling.

1 The Artist

1. A. Ehrenzweig, *The Hidden Order in Art* (London: Weidenfeld and Nicholson, 1967), 50–51. "The third slow movement of the Hammerclavier Sonata still startles me by a sudden twist that breaks into the broad beautiful cantilena and produces a melodic as well as an harmonic rupture. The notebooks tell us that it was not the broad adagio theme but this abrupt transition which Beethoven first noted down. How strange: a transition between melodies not yet existing! The melodies unfold later from this rupture between them. Beethoven never revised the break while he kept on filing and refining the broad melodies. Here we have a good example of an inarticulate disruptive idea which guides and unfolds the large-scale structures."

2. In R. E. M. Harding, *An Anatomy of Inspiration* (New York: Barnes and Noble, 1967), 51.
3. A. Georgotas and R. Cancros, eds., *Depression and Mania* (New York: Elsevier Science Publications, 1988).
4. See P. Rentchnick, *Ces malades qui nous gouvernent* (Paris: Stock, 1976). See also J. J. Schildkraut, et al., "Mind and Mood in Modern Art: Depressive Disorders, Spirituality, and Early Deaths in the Abstract Expressionist Artists of the New York School," *The American Journal of Psychiatry* (April 1994).
5. H. Segal in H. Read, *The Origins of Form in Art* (New York: Horizon Press, 1965), 62.
6. Cited in F. Westley and H. Mintzberg, "Visionary Leadership and Strategic Management," *Strategic Management Journal* 10 (1989).

2 The Craftsman

1. Cited in John Saul, *Voltaire's Bastards* (New York: Penguin Books, 1992), 67.
2. H. A. Sackheim, "Self-deception, Self-esteem, and Depression: The Adaptive Value of Lying to Oneself," in J. Masling, ed., *Empirical Studies of Psychoanalytic Theories* (New Jersey: Erlbaum, 1983).
3. See Laurent Lapierre, "Le Ménagement": ménager, faire le ménage, et se ménager," *Gestion: Revue Internationale de Gestion* (November 1992). Lapierre does not attribute these attitudes to the Craftsman, but I claim it is the Craftsman that he is describing. He agrees. For the most part, the Artist is not close enough to day-to-day events and people to notice who needs to be disciplined.

3 The Technocrat

1. Robertson Davies, *Tempest Tost* (London: Penguin Books, 1980.)
2. There may be more sincere motivations for this interest in art. It may be that art is the one place where, in technocratic eyes, it is legitimate to have feelings. It should go without saying that I am not claiming that all art collectors are Technocrats, but I thought I'd better add this just in case.
3. Goethe, in K. Schnier, "The Function and Origin of Form," *Journal of Aesthetics and Art Criticism* 16 (1957–1958), 98.
4. Platon, *Protagoras* (Paris: Librairie Générale Française, 1993).
5. In technical language, this is *projective identification*—"He's like me, what a great guy!" Or *introjective identification*—"Boy, would I like to be like him."
6. American Psychiatric Association, *Diagnostic and Statistical Manual of Mental Disorders (DSMIII)* (Washington: American Psychiatric Association, 1987).
7. W. G. Dahlstrom, G. S. Welsh, and L. E. Dahlstrom, *An MMPI Handbook,* vol. 1 (Minneapolis: University of Minnesota Press, 1972), 267–271.
8. *Diagnostic and Statistical Manual of Mental Disorders (DSMIII),* op. cit.
9. See S. Kakar, *Frederick Taylor: A Study in Personality and Innovation* (Cambridge: M.I.T. Press, 1970).
10. David Shapiro, *Neurotic Styles* (New York: Basic Books, 1965), 192.
11. L. Hinsie and R. J. Campbell, *Psychiatric Dictionary* (London: Oxford University Press, 1970), 105.

5 The Dreams of the Playwright

1. S. K. Langer, *Mind: An Essay on Human Feeling,* vol. 1 (Baltimore: Johns Hopkins University Press, 1967), 90.

7 Climax: The Triumph of Technical Illusions

1. Alleged report of a work-study engineer circulating in the United Kingdom in the 1950s, cited in William J. Baumol and William G. Bowen, *Performing Arts—The Economic Dilemma* (Cambridge: M.I.T. Press, 1966), 165.

8 Dénouement: The Collapse of Vision

1. Charles Taylor, *The Sources of the Self: The Making of the Modern Identity* (Cambridge: Harvard University Press, 1989), 24.
2. This "strategic" approach to human relations is characteristic of the paranoid personality: Assuming that others are plotting against him, he adopts what seems to him to be a legitimate defensive strategy against them.
3. From *sophist:* paid teacher of philosophy in ancient Greece; reasoner willing to avail himself of fallacies that will help his case. *Sophism:* a false argument meant to deceive; *sophisticate:* spoil the simplicity or purity or naturalness of, corrupt or adulterate or tamper with.
4. Barbara Tuchman, *A Distant Mirror* (New York: Alfred A. Knopf, 1978).

9 Leadership Revisited

1. R. G. Collingwood, *The Principles of Art* (London: Oxford University Press, 1938).
2. Abraham Zaleznik, "Managers and Leaders: Are They Different?" *Harvard Business Review* (May–June 1977).

3. Abraham Zaleznik, *The Managerial Mystique* (New York: Harper and Row, 1989).
4. The Book of Job, The Bible.
5. Alain-Gérard Slama, *L'Angélisme exterminateur* (Paris: Bernard Grasset, 1993).

10 Where There's Smoke There's Apt to Be Fire

1. James Collins and Jerry Porras, *Built to Last* (New York: HarperCollins, 1994).
2. Gene Landrum, *Profiles of Genius* (New York: Prometheus Books, 1993), 213.
3. Kay Jamieson, *Touched with Fire* (New York: Free Press, 1993).
4. Diane Disney-Miller and Pete Martin, *The Story of Walt Disney* (New York: Holt, 1957), 237.
5. Collins and Porras, *Built to Last.*
6. Collins and Porras, *Built to Last,* 180.

11 About Teams

1. Preface, in Otto Rank, *Art and Artist* (New York: Alfred A. Knopf, 1932).
2. J. B. Quinn, "Strategic Change: Logical Incrementalism," *Sloan Management Review* (Fall 1978).
3. Hervé Sérieyx, *Le Big Bang des Organisations* (Paris: Calmann-Lévy, 1993).
4. *La Presse,* Montreal, 12 February 1994.
5. Laurent Lapierre, "Diriger ou pas Diriger: Voilà la question!" *Gestion: Revue Internationale de Gestion* 16, no. 4 (November 1991), 54–55.
6. R. Giuliani, "Freedom Is About Authority," *New York Times,* 20 March 1994.

7. S. Arieti, *Creativity: The Magic Synthesis* (New York: Basic Books, 1976).

12 The Scope of the Problem

1. W. B. Yeats, “The Second Coming,” in Richard J. Finneran, ed., *The Collected Poems of W. B. Yeats* (New York: Simon & Schuster, 1996).
2. Michel Bauer and Benedicte Bertin-Mourot, “Production d’autorité legitime,” *Symposium on Enterprise and Society* (Montreal: L’École des Hautes Études Commerciales 21 August 1995).
3. *English Etymology* (London: Oxford University Press, 1993), 79.
4. A. Glassman, “Cigarette Smoking: Implications for Psychiatric Illness,” *American Journal of Psychiatry* (April 1993).

13 Partial Solutions

1. John Saul, *Voltaire’s Bastards* (New York: Penguin Books, 1992).
2. Antonio Damasio, *Descartes’ Error* (New York: Putnam and Sons, 1994).
3. Charles Taylor, *The Sources of the Self* (Cambridge: Harvard University Press, 1989).
4. Robert McNamara, *In Retrospect* (New York: Times Books, 1995).
5. Alain-Gérard Slama, *L’Angélisme exterminateur* (Paris: Bernard Grasset, 1993).
6. Allan Bloom, *The Closing of the American Mind* (New York: Simon & Schuster, 1987).

7. C. S. Lewis, "On the Reading of Old Books," in *First and Second Things* (London: Collins and Sons, 1985), 27.
8. T. S. Eliot, "Choruses from the Rock."

Appendix B Why Art

1. Herbert Simon, *The New Science of Management Decision* (Englewood Cliffs, N.J.: Prentice-Hall, 1977).
2. Howard Gardner, *The Mind's New Science* (New York: Basic Books, 1985).
3. Ibid., 6.
4. M. E. Porter, *Competitive Strategy* (New York: Free Press, 1980), and *Competitive Advantage* (New York: Free Press, 1985). In a little-noted footnote, Porter admits that his analytic framework can be seen as a substitute for vision (1980:151), but this is a little like saying ball bearings can be a substitute for wheels; they're round but they won't take you very far or very fast.
5. Among others, R. B. Zajonc, "Feeling and Thinking" and "On the Primacy of Affect," *American Psychologist* (February 1980 and February 1984, respectively).
6. A. Isen and K. Daubman, "The Influence of Affect on Categorization," *Journal of Personality and Social Psychology* 47, no. 6 (1984), 1217.
7. David Shapiro, *Neurotic Styles* (New York: Basic Books, 1965), 192.
8. Warren Bennis, *On Becoming a Leader* (New York: Addison-Wesley, 1989). Frances Westley and Henry Mintzberg, "Visionary Leadership and Strategic Management," *Strategic Management Journal* 10 (1989).

9. Herbert Read, *The Origins of Form in Art* (New York: Horizon Press, 1965), 88.

10. Michael Polanyi, *Personal Knowledge* (Chicago: University of Chicago Press, 1958), 18.

11. S. K. Langer, *Feeling and Form* (New York: Scribner's, 1953), 28.

12. Cited in W. O'Flaherty, *Dreams, Illusions and Other Realities* (Chicago: University of Chicago Press, 1984), 8.

13. E. Gombrich, *Art and Illusion* (Princeton: Princeton University Press, 1961).

14. In J. P. Sartre, *Essays in Existentialism* (New York: Citadel Press, 1965).

15. Fiedler in F. Sparshott, *The Theory of the Arts* (Princeton: Princeton University Press, 1982), 626.

16. Lord Brain, *Some Reflections on Genius* (London: Pitman Medical Publishing, 1960).

17. Roger Fry, *Vision and Design* (Harmondsworth: Penguin Books, 1920), p. 29.

18. In V. Tomas, ed., *Creativity in the Arts* (Englewood Cliffs, N.J.: Prentice-Hall, 1964), 47.

19. F. Barron, "Creative Writers," in R. S. Albert, ed., *Genius and Eminence* (New York: Pergamon Press, 1983), 309.

20. In J. Schnier, "The Function and Origin of Form," *Journal of Aesthetics and Art Criticism* (1957–1958), 69.

21. In Tomas, *Creativity,* 40.

22. Ibid., 82.

23. In Sparshott, *Theory of Arts,* 623. Emphasis in the original.

24. J. Bronowski, "The Creative Process," *Scientific American* (September 1958).

25. In Sparshott, *Theory of Arts,* 616.

26. In Lord Brain, *Reflections on Genius.*
27. C. G. Jung, in J. Campbell, ed., *The Portable Jung* (London: Penguin Books, 1971).
28. S. Arieti, *Creativity: The Magic Synthesis* (New York: Basic Books, 1976). M. Milner, *The Suppressed Madness of Sane Men* (London: Tavistock Publications, 1987). A. Ehrenzweig, *The Hidden Order in Art* (London: Weidenfeld and Nicolson, 1967).
29. Rollo May, *The Courage to Create* (New York: Bantam Books, 1975), 62.
30. George Santayana, *Reason in Art* (New York: Dover Publications, 1982), 56.
31. George Santayana, *Le Petit Robert* (Paris: Dictionnaire le Robert, 1990).
32. W. Reich, *Character Analysis* (New York: Orgone Institute, 1949), 199.
33. Polanyi, *Personal Knowledge,* 240.
34. C. G. Jung, in J. Campbell, ed., *The Portable Jung* (London: Penguin Books, 1974), 314.
35. H. Mintzberg, "Crafting Strategy," *Harvard Business Review* (July–August 1987).
36. In Sparshott, *Theory of Arts,* 198.
37. H. Osborne, Preface, in *Oxford Companion to Craft* (London: Oxford University Press, 1975), and "The Aesthetic Concept of Craftsmanship," *British Journal of Aesthetics* (Spring 1977).
38. Sparshott, *Theory of Arts,* 131.
39. V. A. Howard, *Artistry: The Work of Artists* (Indiana: Hackett Publishing, 1982), 35.
40. Polanyi, *Personal Knowledge,* 53.

41. The details of the organization, names, dates, assets, and locations are deliberately vague in order to protect the anonymity of the managers.
42. H. G. Gough, "The Adjective Checklist as a Personality Assessment Research Technique," *Psychological Reports* 6 (1960).

Appendix C Technical Material

1. R. M. Hogarth and S. Makridakis, "Forecasting and Planning: An Evaluation," *Management Science* 27, no. 2 (February 1981). A. Tversky and D. Kahneman, "The Framing of Decisions and the Psychology of Choice," *Science* 211 (January 1981).
2. F. Kerlinger, *Foundations of Behavioral Research* (New York: Holt, Rhinehart and Winston, 1986), 569.
3. Kerlinger, *Foundations.* J. E. Overall and C. J. Klett, *Applied Multivariate Analysis* (New York: McGraw-Hill, 1972).
4. Overall and Klett, *Multivariate Analysis.* D. Miller and P. Friesen, "Archetypes of Organizational Transition," *Administrative Science Quarterly* (June 1980).
5. Kerlinger, *Foundations,* 581.
6. Ibid., 582.
7. J.-P. Fenelon, *Qu'est-ce que c'est l'analyse des données?* (Paris: Lefonen, 1981). L. Lebart, A. Morineau, and K. M. Warwick, *Multivariate Descriptive Statistical Analysis* (New York: John Wiley and Sons, 1984).
8. J. P. Benzecri et al. *L'analyse des Correspondences* 2 (Paris: Dunod, 1979).
9. Lebart, Morineau, and Warwick, *Multivariate Statistical Analysis,* 30.

INDEX